Biomolecular Free Radical Toxicity: Causes and Prevention

Biomolecular Free Radical Toxicity: Causes and Prevention

HELEN WISEMAN
Nutrition, Food and Health Research Centre, Department of Nutrition and Dietetics, King's College London, Franklin-Wilkins Building, 150 Stamford Street, London SE1 8WA, UK
PETER GOLDFARB
Molecular Toxicology Group, School of Biological Sciences (SBS), University of Surrey, Guildford, Surrey GU2 7XH, UK
TIM RIDGWAY
Wise Associates at School of Biological Sciences (SBS), University of Surrey, Guildford, Surrey GU2 7XH, UK
ALAN WISEMAN
Molecular Toxicology Group, School of Biological Sciences (SBS), University of Surrey, Guildford, Surrey GU2 7XH, UK

With contributions from:
Sek C. Chow, *Centre for Mechanisms of Human Toxicity, University of Leicester, Leicester, UK*
Barbara R. Evans, *Biotechnology, Oak Ridge National Laboratory, Tennessee, 37831-6194, USA*
Richard J. Hambly, *Department Medical Sciences, University of Bath, Bath, BA2 7AY*
Costas Ioannides, *Molecular Toxicology Group, SBS, University of Surrey Guildford, Surrey, GU5 7XH*
George E. N. Kass, *Molecular Toxicology Group, SBS, University of Surrey, Guildford, Surrey GU2 7XH*
Sten Orrenius, *Karolinska Institute, Stockholm, Sweden*
Jonathan Woodward, *Biotechnology, Oak Ridge National Laboratory, Tennessee, 37831-6194, USA*

JOHN WILEY & SONS, LTD
Chichester · New York · Weinheim · Brisbane · Singapore · Toronto

Commissioned in the UK on behalf of John Wiley & Sons, Ltd
by Medi-Tech Publications, Storrington, West Sussex RH20 4HH, UK.

Other Wiley Editorial Offices

John Wiley & Sons, Inc., 605 Third Avenue,
New York, NY 10158-0012, USA

VCH Verlagsgesellschaft mbH, Pappelallee 3,
D-69469 Weinheim, Germany

Jacaranda Wiley Ltd, 33 Park Road, Milton,
Queensland 4046, Australia

John Wiley & Sons (Asia) Pte Ltd, 2 Clementi Loop #02-01
Jin Xing Distripark, Singapore 129809

John Wiley & Sons (Canada) Ltd, 22 Worcester Road,
Rexdale, Ontario M9W 1L1, Canada

Library of Congress Cataloging-in-Publication Data
Biomolecular free radical toxicity : causes and prevention / Helen Wiseman . . . [et al];
with contributions from Sek C. Chow . . . [et al.].
    p. cm.
  Includes bibliographical references and index.
  ISBN 0-471-49076-8 (alk. paper)
  1. Free radicals (Chemistry)–Pathophysiology. 2. Pathology, Molecular. 3.
Cytochrome P-450. 4. Molecular toxicology. I. Wiseman, Helen.

RB170.B568 2000
616.07–dc21   00-042266

British Library Cataloguing in Publication Data

A catalogue record for this book is available from the British Library

ISBN 0 471 49076 8

Typeset in 10/12 pt Palatino by Dobbie Typesetting Limited, Tavistock, Devon
Printed and bound in Great Britain by Biddles Ltd, Guildford and King's Lynn
This book is printed on acid-free paper responsibly manufactured from sustainable forestry,
for which at least two trees are planted for each one used for paper production.

Contents

Preface

This book presents an account of biomolecular injury that recognises the subtle nature of some damage to cellular macromolecules. Biomolecular injury may manifest rapidly, or slowly if initially confined to genomic DNA. Long-term onset of diseases as a result of oxidative biomolecular damage and subsequent injury are now well recognised as the cause of diseases associated with mutagenesis (DNA injury – see Chapter 2) and in some cases consequent carcinogenesis (see Chapter 4). Attention is given therefore to injury to DNA and proteins (see Chapter 3), in addition to membrane lipids and lipoproteins (see Chapter 1). Activation of foreign chemicals by cytochromes P-450 can give rise to mutagens and carcinogens (see Chapter 4). The remarkable field of apoptosis (programmed cell death) is presented (see Chapter 5) in its rapidly expanding importance to cellular events. The use of antioxidants to protect against oxidative biomolecular damage and subsequent injury is a theme widely explored in the book (see Chapters 1–3). A fully systematic research approach in the future would be expected to identify the cooperative interactions of cellular antioxidant systems raised in opposition to pro-oxidant phenomena. Herein lies the route to designer-antioxidants, of use in the combat of oxidative stress through metabolism of components of foodstuffs. Prevention of biomolecular injury is likely to be one of a number of mechanisms by which phytoestrogens, including the isoflavones found in soya foods, also other flavonoids and the drug tamoxifen and related compounds may protect against diseases including hormone-dependent cancer and heart disease (see Chapter 6). The related issue of the ecotoxicity of environmental oestrogens and oestrogen mimics, particularly in relation to strategies for their removal from the environment is also considered (see Chapter 7).

Therefore, this book draws together, through its individual chapters, these diverse fields of knowledge and discovery, and provides interpretations of data, of major importance in biochemistry, and in toxicology, through the understanding of the causation of biomolecular injury and of the possible routes to its prevention.

Acknowledgement

The Authors express their gratitude to Mrs Margaret Whatley for excellent word-processing.

The Authors: July 2000

Macromolecular Mechanisms of Injury and the Role of Antioxidants

1 Membrane Lipid and Lipoprotein Injury: Prevention by Antioxidants

HELEN WISEMAN[1] and **TIM RIDGWAY**[2]

[1]*Nutrition, Food and Health Research Centre, King's College London, Franklin-Wilkins Building, 150 Stamford Street, London SE1 8WA, UK*

[2]*Wise Associates, School of Biological Sciences, University of Surrey, Guildford, Surrey GU2 7XH, UK*

Membrane structural and functional features

The plasma (cytoplasmic) membrane surrounds the cell, and with other membranes form a continuous intracellular surface (endoplasmic reticulum) and form other intracellular organelles such as mitochondria (Gurr and Harwood 1991). Lipids in animal cell membranes are mainly phospholipids based on glycerol, such as phosphatidylserine, phosphatidylethanolamine, phosphatidylcholine and phosphatidylinositol, each with variable fatty acid side chains (Gurr and Harwood 1991). The phospholipid bilayer of all membranes and the presence of a wide range of different proteins confers a great diversity of function. Cholesterol is found in large amounts in plasma membranes (often equimolar with the phospholipid), whereas endoplasmic reticulum, mitochondrial and nuclear membranes have a low cholesterol content (Gurr and Harwood 1991). In the basic model for membrane structure, extrinsic proteins are loosely adsorbed to the lipid bilayer, while the intrinsic proteins are embedded in the bilayer and can traverse the whole bilayer (Gurr and Harwood 1991; Peck 1994). Lipid and protein molecules can rapidly diffuse laterally, whereas migration from one bilayer to another (flip-flop) is rare and slow and this preserves the asymmetry of natural membranes (Zachowski 1993; Williamson and Schlegel 1994). Membrane fluidity encompasses this lateral diffusion possibility and is dependent on the presence of unsaturated and polyunsaturated fatty acid side chains in the membrane lipids, which lowers the melting (phase) transition temperature of the system (Gurr and Harwood 1991; New 1992). Increasing the cholesterol (and the saturated fatty acid) content of a membrane decreases membrane fluidity. Cholesterol has a rigidifying effect on the membrane by inhibiting the overall flexing motion of the hydrocarbon chains. Membrane fluidity has a strong influence on membrane conformation and thus activity of membrane-associated enzymes and receptors involved in a wide range of functions, from cell signalling to nutrient uptake (Peck 1994).

Oxidative biomolecular injury to membrane lipids

Lipid peroxidation is a free radical-mediated chain reaction (which can be initiated by the hydroxyl radical) that attacks polyunsaturated fatty acids in membranes and plasma lipoprotein particles resulting in oxidative damage (Halliwell and Gutteridge 1999; Halliwell 1995a,b). Free radicals are any species capable of an independent existence that contain one or more unpaired electrons (Halliwell and Gutteridge 1999; Halliwell 1995a,b; Knight 1998). Reactive oxygen species (ROS) refers to oxygen-centred radicals such as superoxide and the hydroxyl radical, and also to hydrogen peroxide, singlet oxygen, hypocholorous acid and ozone. This group thus includes dangerous, non-radical derivatives of oxygen (Wiseman and Halliwell 1996). Reactive nitrogen species (RNS) can be derived from nitric oxide and include the damaging species peroxynitrite (Darley-Usmar *et al.* 1995). Free radicals (Diplock *et al.* 1994; Cerutti, 1994; Witzum 1994; Knight 1998; Cross *et al.* 1998) are produced as the by-products of normal metabolism (superoxide, the hydroxyl radical and nitric oxide and ROS such as hydrogen peroxide are all formed *in vivo*) and have been implicated in over 100 diseases including cardiovascular disease and cancer. However, in many cases they may not be the major cause of disease but rather a complicating component of the underlying disease pathology, leading to lipid peroxidation as a consequence rather than a cause of cell injury (see Tables 1.1 and 1.2 and Figures 1.1, 1.2 and 1.3).

Membrane lipid peroxidation damages membrane proteins through consequent free radical attack (Dean *et al.* 1993; Stadtman 1993). Lipid hydroperoxides can be readily decomposed by traces of transition metal ions to produce the free radical intermediates of lipid peroxidation capable of propagating the chain reaction. ROS/RNS-mediated damage to DNA bases is implicated in mutation and tumorigenesis (Cerutti 1994; Wiseman and Halliwell 1996; Wiseman 1996; Poulsen *et al.* 1998). Oxidative DNA damage includes OH˙-mediated modification of DNA bases. ROS/RNS induced mutations could result not only from direct DNA damage but also indirectly as a consequence of oxidative damage to membranes (Wiseman and Halliwell 1996; Wiseman 1996). This attack on lipids in membranes can initiate the process of lipid peroxidation and lipid peroxides formed as a result of this oxidative membrane damage can subsequently decompose to mutagenic carbonyl products (Cheeseman 1993). Lipid peroxidation may have a role in human breast cancer risk: urinary excretion of the mutagen malondialdehyde has been shown to be approximately double in women with mammographic displasia (high risk) than in women without these changes (Boyd and McGuire 1991).

Protein modifications include oxidation of thiol groups and the generation of carbonyl derivatives of amino acid residues (Oliver *et al.* 1987). Nitrosothiols and nitrotyrosine residues may occur as a result of RNS attack (Darley-Usmar *et al.* 1995). Monosaccharide autooxidation is a transition metal-catalysed process that generates the highly reactive products (hydrogen peroxide and ketoaldehydes) that can modify proteins (Wolff and Dean 1987). This process could thus cause damage to membranes (an excess of glucose and other monosaccharides in the diet may thus be deleterious).

The generation of ROS (and RNS) should be in balance with antioxidant defences. An antioxidant is 'any substance that, when present at low concentrations compared to those of an oxidizable substrate, significantly delays or prevents oxidation of that substrate' (Halliwell 1995a). 'Oxidizable substrates' include DNA, lipids, proteins (including of membranes and lipoproteins) and carbohydrates (Halliwell 1995a). Imbalance between generation of free radicals and antioxidant defences is referred to as oxidative stress (Halliwell and Gutteridge 1999). Oxidative stress can be caused by the following:

1. Increased ROS formation, for example, caused by toxic chemicals and drugs (including cytochrome P-450 dependent futile-cycling resulting in the formation of superoxide radicals as by-products) and at sites of inflammation (as a result of the phagocyte oxidative burst).
2. Depletion of antioxidant levels, for example, malnutrition lowers antioxidant vitamin and glutathione levels. A slow general accumulation of oxidative damage contributes to the ageing process and age-related diseases such as cancer (Ames 1989). Mitochondria suffer oxidative stress because their energy production is also associated with the production of ROS. Oxidative damage to mitochondria includes damage to the mitochondrial membrane (increased levels of lipid peroxidation products have been found in mitochondria exposed to oxidative stress). This has been implicated in neurodegenerative disorders such as Parkinson's disease and ageing (Shigenaga *et al.* 1994; Jenner 1994).

Membrane lipid peroxidation is often measured in microsomes or liposomes. Microsomes, a heterogeneous mixture of vesicles derived from both endoplasmic reticulum and plasma membranes, are used as an *in vitro* test system to assess the ability of a wide range of drugs and dietary components to protect (as antioxidants) against membrane lipid peroxidation (Halliwell and Gutteridge 1999). Liposomes are used extensively as a model membrane system for studying the influence of dietary components and drugs on membrane lipid peroxidation *in vitro* (Wiseman 1996).

Liposomes are artificial lipid structures, and are made by shaking or sonicating phospholipids in aqueous suspension (New 1992).

The most extensively used method is probably the thiobarbituric acid (TBA) test. The test sample is heated with TBA at low pH and the absorbance of a pink chromogen presumed to be a $(TBA)_2$–malondialdehyde (MDA) adduct (although the term TBARS: thiobarbituric acid reactive substances is frequently used) is measured at 532 mm. Although the TBA test is adequate for measuring lipid peroxidation in defined membrane systems such as microsomes and liposomes, its application to body fluids has many problems relating to its lack of specificity. A modified TBA test has been developed that avoids many of the artefacts resulting from the reaction of TBA with other body-fluid constituents to give different chromogens and uses high-performance liquid chromatography to separate the authentic $(TBA)_2$–MDA adduct from other chromogens absorbing at 532 mm (Halliwell and Chirico 1993).

It appears, however, that measurement of plasma levels of F_2-isoprostanes, specific end products of the peroxidation of arachidonic acid residues, by mass spectrometry may currently be the biomarker of choice for lipid peroxidation in the human body (Pratico *et al.* 1998; Roberts and Morrow 1997; David *et al.* 1997; Morrow *et al.* 1995; Basu 1998; Roberts *et al.* 1998). Wide variations in F_2-isoprostane levels have been found in the body fluids of healthy subjects, which could depend in part on diet (Pratico *et al.* 1998; Roberts and Morrow 1997). However, these may also include other factors such as endogenous antioxidant defences and rate of isoprostane metabolism (Basu 1998). Furthermore, some healthy human subjects appear to show increased rates of lipid peroxidation compared to others even on comparable diets and could be at increased risk from diseases involving lipid peroxidation such as atherosclerosis and cancer.

Membrane fatty acids in oxidative biomolecular injury

Fatty acids in the diet such as the polyunsaturated n-3 fatty acids have been reported to increase human erythrocyte membrane fluidity (Berlin *et al.* 1992). This increase in membrane fluidity may result from the esterification of *cis* polyunsaturated fatty acids to the membrane phospholipids (Berlin *et al.* 1992). In contrast, dietary substitution of saturated fatty acids with *cis* or *trans* monounsaturated fatty acids has been shown to have no influence on human erythrocyte membrane fluidity (Berlin *et al.* 1994). Olive oil is rich in monounsaturated oleic acid compared to blackcurrant oil, which is rich in linoleic and 18C-polyunsaturated fatty acids but deficient in oleic acid. Investigation of brain and heart membranes taken from rats fed diets rich in either olive oil or blackcurrant oil suggest that the fatty acid composition of the olive oil-based diet may predispose towards lower membrane fluidity

Table 1.1.   Worst-outcome versus best-outcome scenarios: toxic consequences of biomolecular injury

Worst outcome	Best outcome
Antioxidant failure	Antioxidant success, e.g. superoxide dismutase
Biomolecular injury fails to respond to repair (or therapy)	Biomolecular injury repair
Mutagenesis: with carcinogenesis in animals and plants	Undetectable mutagenesis without carcinogenesis in animals and plants
Morbidity and mortality	Health maintained or restored
Teratogenesis	No teratogenesis
Inborn errors of metabolism in future generations of offspring (descendants)	No inborn errors of metabolism in future generations of offspring (descendants)

than the blackcurrant oil (Barzanti *et al.* 1995). Furthermore, less membrane lipid peroxidation was observed with the olive oil diet (Barzanti *et al.* 1995), possibly because the decreased levels of polyunsaturated fatty acids in the membranes make them less susceptible to free radical attack (see below). Increased susceptibility to membrane lipid peroxidation has been observed in the livers of rats fed a high fish oil diet rich in polyunsaturated fatty acids (Saito and Nakatsugawa 1994). Dietary polyunsaturated fatty acids (cod liver oil rich in n-3 fatty acids was used) have been found to enhance lipid peroxide formation in rat liver, because of increased membrane lipid peroxidation (Reddy and Lokesh 1994). There may be a requirement for an increased antioxidant intake to accompany an increased polyunsaturated fatty acid intake to obtain the suggested beneficial influence of poly-unsaturated fatty acids against heart disease. In healthy volunteers given catechin-type flavonoids in green tea for four weeks, an overall protection against modification of red blood cell membrane polyunsaturated fatty acids was observed, probably arising from the catechins sparing the endogenous antioxidants vitamin E and β-carotene (Pietta and Simonetti 1998).

Prevention of oxidative biomolecular injury by antioxidants in the diet (see Chapters 2 and 3)

Protection by vitamins

α-Tocopherol inhibits lipid peroxidation in microsomal and liposomal membrane systems (Kagan *et al.* 1990a). The membrane antioxidant action of α-tocopherol enables protection of cultured human endothelial

Table 1.2. Modes of biomolecular injury

Biomolecule	Modes of injury
DNA and RNA	Main-chain cleavage
	Apurination
	Base conversion
	Damage to pentose
	Hydrogen bonding misdirected
	Nucleotide deletion
Proteins and enzymes	Main-chain cleavage
	Main-chain–CH_2 to $C=O$
	Amino-acid residue damage
	Thiols SH to disulphides SS
	(possibly reversible)
	Loss of biological function
	(denaturation possibly reversible
	by change in configuration or
	conformation)
Phospholipids in membranes	Oxidation and fragmentation caused
	by free-radical species

cells against linoleic acid hydroperoxide-induced cytotoxicity, due to hydroperoxide-induced lipid peroxidation (Kaneko *et al.* 1994). Vitamin E (tocopherols) fed to chickens protected against membrane lipid peroxidation in chicken liver slices (Pellett *et al.* 1994). Ethanol-induced oxidative stress can be partly prevented by vitamin E supplementation. The role of chronic alcohol consumption in promoting oxidative stress in the liver, expressed as increased lipid peroxidation of hepatic membranes is well known and is thought to be potentiated by a fat-rich diet and/or xenobiotics (Nordmann 1994). Vitamin E is thought to act as a chain-breaking antioxidant by donating a hydrogen atom from the phenolic hydroxyl group present in the 6-hydroxychromane ring to the chain-propagating lipid peroxyl and alkoxyl radical intermediates of lipid peroxidation, thus terminating the chain reaction and producing lipid hydroperoxides and the tocopheroxyl radical (Morrissey *et al.* 1994; Dutta-Roy *et al.* 1994). An alternative pathway whereby α-tocopherol also exerts a protective effect by converting the peroxyl radical into the parent unoxidized lipid molecule with the concomitant formation of superoxide, has been demonstrated in a liposomal model membrane system (Dmitriev *et al.* 1994).

In addition to the chain-breaking antioxidant action of vitamin E, which is observed as a characteristic time course, the test curves reach the same level as the control curve as the chain-breaking antioxidant is consumed (Wiseman and Halliwell 1996), a structural membrane antioxidant action mediated by the hydrocarbon chain of vitamin E through decreased membrane fluidity has also been demonstrated. The membrane sterol cholesterol is also proposed to act as a structural antioxidant via interaction between its hydrophobic rings and the saturated, monounsaturated and polyunsaturated residues of phospholipid fatty acid side chains, and is not a chain-breaking antioxidant despite possessing a potentially donatable hydrogen atom (Halliwell and Gutteridge 1999, Wiseman 1994a). Indeed, measurements of membrane fluidity have shown that cholesterol and vitamin E have a rigidifying effect and notable decreases in membrane fluidity have also been shown for oestrogens and for the antioestrogen drug tamoxifen, widely used in the treatment of breast cancer and increasingly proposed for its prevention (see Chapter 6) (Clarke et al. 1990). A similar mechanism of action has thus been suggested for the antioxidant action of tamoxifen (it does not possess a donatable hydrogen atom for chain-breaking antioxidant action) and also 4-hydroxytamoxifen and oestrogens and vitamin D (each possesses a potentially donatable hydrogen atom, but time-course data suggests they are not predominantly chain-breaking antioxidants) (Sugioka et al. 1987; Wiseman et al. 1990; Wiseman 1994a,b; Wiseman 1993; Ruiz-Larrea et al. 1994; Lacort et al. 1995). A structural membrane antioxidant action is proposed for these compounds on the basis of their ability to mimic the membrane stabilisation against lipid peroxidation demonstrated by the endogenous membrane sterol cholesterol, to which all these compounds show some degree of structural similarity (Wiseman 1994a). In addition to their chain-breaking antioxidant activity (and ability to bind iron and copper ions, see below), flavonoids and particularly isoflavonoids, which demonstrates some structural similarity to oestrogens and tamoxifen may also act as membrane antioxidants by membrane stabilising mechanisms similar to those proposed for cholesterol, oestrogen and tamoxifen (Wiseman 1994a).

When vitamin C was fed to guinea pigs (for a period of five weeks), a dose of 660 mg/kg diet was found to be optimal for reducing lipid peroxidation in the liver (Barja et al. 1994). Vitamin C at this dose (660 mg/ kg diet) also reduced TBARS formation when liver microsomal membranes were peroxidised in vitro. This indicates that a diet supplying 40 times more vitamin C than is required to prevent scurvy, protected membranes and other cellular components against oxidative damage (Barja et al. 1994). Vitamin C concentrations have been found to be lower in the breast milk of smokers, which may aggravate the peroxidation problems (maternal smoking contributes to peroxidation events in newborn infants) of their newborn infants (Ortega et al. 1998). Studies have also indicated that increased antioxidant intake is associated with decreased coronary disease risk and a low plasma ascorbic acid concentration has been shown independently to predict the presence of an unstable coronary syndrome (Vita *et al.* 1998). Furthermore, increased ascorbic acid intake is associated with decreased risk of coronary heart disease and stroke (Simon *et al.* 1998).

Ascorbic acid can scavenge free radicals in the aqueous phases of cells and in the circulatory system and protects membranes and lipoprotein particles from oxidative damage by regenerating the antioxidant form of vitamin E (Buettner 1993; Beyer 1994). Ubiquinol-10 (the reduced form of ubiquinone-10 or coenzyme Q-10 found in green leafy vegetables and sold as a health supplement) is another effective membrane antioxidant and its protective action has been demonstrated in liposomal membranes (Frei *et al.* 1990). Ubiquinol-10 is also thought to interact with vitamin E to regenerate its antioxidant form (Beyer 1994; Kagan *et al.* 1990b).

β-Carotene acts as a radical quenching antioxidant at low oxygen tensions (Burton and Ingold 1984), (it is particularly effective against singlet oxygen) and may be a membrane antioxidant at the relatively low oxygen tension at intracellular membranes. Indeed, β-carotene supplementation has been reported to reduce lipid peroxidation *in vivo* (Allard *et al.* 1994). Carotenoid mixtures have been shown to protect liposomes against lipid peroxidation measured as TBARS: mixtures were more effective than the single compounds and this synergistic effect was most pronounced when lutein or lycopene were present (Stahl *et al.* 1998).

In the Penn State Young Women's Health Study, lower weight, lower body mass index and higher circulating α-tocopherol and β-carotene concentrations and higher fruit, carbohydrate and fibre intakes were found in the first quintile by percentage body fat compared to the fifth quintile (Lloyd *et al.* 1998). Antioxidant vitamin (β-carotene and α-tocopherol) concentrations in the plasma of obese girls in Japan were relatively lower than in normal controls (Kuno *et al.* 1998). Increases in plasma antioxidant capacity (measured as oxygen radical absorbance capacity (ORAC)) and plasma α-tocopherol concentration have been reported following consumption by humans of controlled diets high in fruit and vegetables and the increase in ORAC could not be explained by the increase in plasma α-tocopherol concentration, suggesting that other antioxidants derived from fruit and vegetables are also important (Cao *et al.* 1998).

Vitamin D is unique among vitamins in that its requirements can be met both from the diet and from skin photobiosynthesis (Anderson and Toverud 1994). Vitamin D is a membrane antioxidant in that it inhibits lipid peroxidation in the liposomal model membrane system (Wiseman 1993), and may also be an antioxidant *in vivo*. The similar antioxidant effectiveness of its precursor in the skin 7-dehydrocholesterol (Wiseman 1993), is of interest.

In addition to their interaction with free radicals, antioxidants can also exert their effects by altering the cellular redox potential. The sulphur-containing

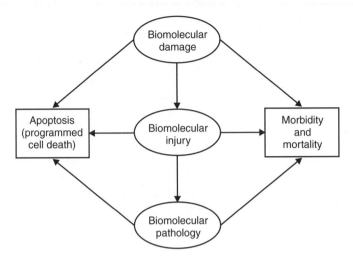

Figure 1.1.   Biomolecular-web causing cellular malaise

antioxidant N-acetylcysteine (but not the chain-breaking antioxidant vitamin E) can induce p53-dependent apoptosis in transformed cell lines and primary cultures (but not normal cells) by elevating p53 (tumour suppressor gene) expression post-transcriptionally by increasing the rate of p53 mRNA translation (Liu *et al*. 1998).

Protection by dietary flavonoids, phytoestrogens and novel flavonoids

Flavonoids including the flavonols quercetin, kaempferol and myricetin (found in onions and tea), and the catechins (flavanols) such as catechin, epicatechin, epigallocatechin and epigallocatechin gallate (found in tea) are polyphenolic compounds found in many foods of plant origin (Cook and Saman 1996; Hertog and Hollman 1996; Peterson and Dwyer 1998). Isoflavonoid phytoestrogens include the soya isoflavones genistein and daidzein and equol the isoflavan metabolite of daidzein (see Figure 1.4).

In a study of the Dutch diet (Hertog *et al*. 1993), tea, onions and apples appeared to provide some protection from coronary heart disease, and this protective effect has been suggested. Many plant secondary metabolites, including the flavonoids, display antioxidant activity (Bors and Saran 1987; Rice-Evans *et al*. 1995, 1996; Cook and Samman 1996). Although it should be noted that phenolic compounds can also display pro-oxidant activity under certain conditions (Decker 1997).

Flavonoids used as additives may provide technological functions, including the prevention of rancidity of fats. The antioxidant properties of such flavonoid-derived additives could provide health benefits, by protecting against the oxidative damage implicated in many disease states (Wiseman and Halliwell 1996). Indeed, a number of approaches including a greater understanding of the role of specific nutrients in foods in disease prevention and treatment, together with appropriate biotechnology have led to the availability of increasing numbers of potential nutritional products with medical and health benefits termed functional foods (also described as designer foods, pharmafoods and nutraceuticals) (Goldberg 1994). Consumption of food and beverages such as apples and tea, rich in antioxidant flavonoids (including quercetin) may be inversely correlated with death from heart disease in the elderly (Hertog et al. 1993; Hertog and Hollman 1996). New nature-identical food additives, particularly flavonoid derivatives with both health beneficial antioxidant and colourant properties, could, therefore, have potential as components of functional foods.

Dietary flavanoid and phytoestrogen antioxidants

The structure–function relationships of flavonoids in relation to antioxidant activity is well characterised (Bors and Saran 1987; Rice-Evans et al. 1995, 1996; Cook and Samman 1996). One of the main features is the formation of o-diphenolic structures which readily take part in redox coupled reactions. The relative potencies of flavonoids as antioxidants is governed by a set of structure–function relationships: in general optimum antioxidant activity is associated with multiple phenolic groups, a double bond in C2–C3 of the C ring, a carbonyl group at C4 of the C-ring and free C3 (C-ring) and C5 (A-ring) hydroxy groups. Structure–activity relationships have been determined for the antioxidant activities of a series of flavonoids including the isoflavone phytoestrogens using a liposomal test system. The presence of hydroxyl substituents on the flavonoid nucleus enhanced activity, by contrast, substitution by methoxy groups diminished antioxidant activity and substitution patterns on the B-ring were of particular importance to antioxidant activity (Arora et al. 1998). In addition, all the flavonoids tested displayed greater antioxidant efficacies against metal-ion-induced peroxidations than peroxyl-radical-induced peroxidation, indicating the likely importance of metal chelation in the antioxidant activity of these compounds. Furthermore, the interaction of flavonoids (quercetin, kaempferol, rutin and luteolin were studied) with Cu^{2+} ions (chelate formation and modification through oxidation) has been shown to be structurally dependent, which has important implications for their antioxidant properties, for example, for Cu^{2+}-chelate formation the ortho-3′,4,′-dihydroxy substitution in the B-ring was found to be important thus influencing antioxidant activity, and the presence of a 3-hydroxy group enhanced the Cu^{2+}-dependent oxidation of quercetin and kaempferol compared to luteolin and rutin which lack this structural feature (Brown et al. 1998).

The flavonols quercetin and myricetin have been observed to inhibit membrane lipid peroxidation (Laughton *et al*. 1989; Terao *et al*. 1994; Arora *et al*. 1998). The isoflavones genistein and daidzein, and in particular the daidzein metabolite equol, have been reported to display antioxidant properties *in vitro*, predominantly against oxidative damage to membrane lipids and lipoprotein particles (see below) (Ruiz-Larrea *et al*. 1997; Kapiotis *et al*. 1997; Arora *et al*. 1998) and also against oxidative DNA damage (see below). Protection against membrane lipid peroxidation is also observed for endogenous oestrogens such as 17β-oestradiol (Sugioka *et al*. 1987; Wiseman *et al*. 1990; Wiseman 1994a,b; Ruiz-Larrea 1994) catechol oestrogens (Lacort *et al*. 1995) and tamoxifen (Wiseman *et al*. 1990; Wiseman 1994a,b).

Pulse radiolysis of aqueous solutions provides data on the spectral, acid–base and redox properties of phenoxyl radicals derived from 3,4,-dihydroxybenzene derivatives and selected flavonoids: favourable reduction potentials of the phenoxyl radicals suggests that flavonoids may act as efficient antioxidants of alkylperoxyl and superoxide/hydroperoxyl radicals (Jovanovic *et al*. 1994). Furthermore, the antioxidant activities of several flavonoids, including the catechins, as both scavengers of radicals in the aqueous phase and against lipid peroxyl radicals have been measured (Salah *et al*. 1995). Kaempferol-3-*O*-galactoside protects mice against bromobenzene-induced hepatic lipid peroxidation, and flavonoids fed to mice have protective effects against gamma-ray-induced oxidative damage, with luteolin being the most effective (Shimoi *et al*. 1994).

The antioxidant activity of flavonoids may also prevent the damaging action of lipid peroxides generated by activated platelets on endothelial nitric oxide and prostacyclin. They both inhibit platelet aggregation and have vasodilatory activity, and flavonoid binding to platelet membranes may inhibit the interaction of activated platelets with vascular endothelium (Hertog and Hollman 1996). Quercetin displays potent antithrombotic effects: it inhibits thrombin and ADP-induced platelet aggregation *in vitro* and this may be through inhibition of phospholipase C activity rather than through inhibition of thromboxane synthesis (Hertog and Hollman 1996).

The food flavouring agent vanillin inhibited the oxidative degradation of linoleic acid in cereals (Buri *et al*. 1989), and carnosol and carnosic acid, which are constituents of one of the active ingredients in rosemary extract, are membrane antioxidants in microsomal and liposomal membranes (Aruoma *et al*. 1992). Caffeic acid (found in coffee) inhibited membrane lipid peroxidation in *in vitro* test systems and coffee itself strongly protected against the mutagenicity and cytotoxicity of the oxidant *t*-butylhydroperoxide (Stadler *et al*. 1994). Curcumin and eugenol, found in spices, lowered the levels of lipid peroxides in the rat (Reddy and Lokesh 1994).

The total antioxidant capacity of fruits and fruit juices (rich in flavonoids and antioxidant vitamins) has been measured (Miller *et al*. 1995; Wang *et al*.

1996; Miller and Rice-Evans 1997; Eberhardt *et al.* 2000). On a dry weight basis the order was strawberry > plum > orange > pink grapefruit > tomato > kiwi fruit > red grape > white grape > apple > melon > pear > banana (Wang *et al.* 1996). Based on the wet weight of fruits (edible portion) strawberry had the greatest antioxidant capacity and the overall order of potency was strawberry > plum > orange > red grape > kiwi fruit > pink grapefruit > white grape > banana > apple > tomato > pear > melon (Wang *et al.* 1996). The antioxidant capacity of these fruits was mostly provided by the juice fractions with the fruit pulp contributing less than 10%. The commercial fruit juices had an order of potency of grape juice > grapefruit juice > tomato juice > orange juice > apple juice (Wang *et al.* 1996). The effectiveness of the flavonoids and other phenols in red wines compared to white wines has been assessed via the compilation of an antioxidant index (Vinson and Hontz 1995).

There is potential for improving the antioxidant capacity of apple fruit and apple juice by modifying levels both of antioxidant flavonoids such as quercetin and antioxidant vitamins such as vitamin C. The total antioxidant activity of apple juice has been used as a biomarker of the deterioration of apple juice upon storage (Miller *et al.* 1995). In 'longlife apple' juice vitamin C represented a minor fraction of the total antioxidant capacity compared to chlorogenic acid and phloridzin, which were the major antioxidants (Miller *et al.* 1995). This suggests that the enhancement of antioxidant activity by improving the antioxidant properties of phloridzin would be a desirable goal in food biotechnology. Evidently though, prior to commercial use, extensive toxicity assessment would have to be made.

Novel phloridzin derivatives and their potential applications

Phloridzin is a flavonoid (dihydrochalcone) largely restricted to apple (*Malus* sp.). Phloridzin is found in small quantities in the mature apple fruit, but in young apple leaves and twigs accounts for up to 10% of dry weight. Although it is accepted that aromatic compounds in plants are synthesised by one of the three major pathways (acetate–mevalonate, acetate–malonate or shikimic acid pathways), some aromatic compounds such as the flavonoids, together with the stilbenes and the xanthones, are biosynthesised by a route that involves both the acetate–malonate and the shikimic acid pathways. The A-ring of the flavonoid ring structure is derived from three condensed acetate units, while the B-ring and three carbons of the central ring are by contrast derived from cinnamic acid, which results from the deamination of phenylalanine (not tyrosine) by phenylalanine ammonium lyase (PAL). The enzyme chalcone synthase (CHS) catalyses the addition of the malonyl Co A units to the cinnamic acid derivative p-coumaric acid to form chalcones such as naringenin chalcone. Naturally occurring chalcone derivatives are hydroxylated and the hydroxyl groups are often found to be glucosylated or methylated (the pattern of cyclisation). Chalcones are coloured compounds and reduction of the α,β-unsaturated bond produces colourless dihydrochalcones (note: terminology then reverses the A- and B-ring) such as phloridzin. These are quite rare, and phloridzin and its aglycone phloretin are the best characterised of these compounds (Mann 1992).

Antioxidant activity in the standard lard (i.e. prevention of rancidity) test system used in food science, showed that an o-diphenolic structure as part of the A-ring of a dihydrochalcone (e.g. $2',4',6',3,4$-pentahydroxydihydrochalcone: 3-hydroxyphloretin) gives a compound that is more effective as an antioxidant than when the o-diphenolic is attached to either a pyran ring (e.g. as in flavanones, flavones or flavanols) or to an unsaturated $\alpha-\beta$ bond (e.g. as in chalcones) (Dziedzic et al. 1983, 1985). The antioxidant properties of 3-hydroxyphloretin suggest its potential use as a food additive antioxidant. Butein, for example, which is a tetrahydroxychalcone and hence a less potent antioxidant than 3-hydroxyphloretin, which is a pentahydroxydihydrochalcone, was found to be six times more effective in preventing fat rancidity than the commercial food antioxidant BHT (Dziedzic 1983): thus 3-hydroxyphloretin would be expected to be at least six times more potent than BHT and may be preferable as a food additive.

3-Hydroxyphloretin is a particularly effective inhibitor of liposomal lipid peroxidation (Ridgway et al. 1996; Ridgway et al. 1997) with an IC_{50} value (concentration required to inhibit lipid peroxidation by 50%) in the low μM range, which was similar to that of quercetin: the most potent flavonoid tested in this system. Phloretin is also a good inhibitor of lipid peroxidation. 3-Hydroxyphloridzin was an effective inhibitor of lipid peroxidation. Phloridzin itself was a poor antioxidant in the liposomal system. These results indicate that although phloridzin itself is not a good inhibitor of lipid peroxidation in the liposomal model membrane system, its derivatives are much more effective. Phloridzin has antioxidant properties in an aqueous-based system for measurement of antioxidant capacity (Miller et al. 1995). Its lack of ability in the ox-brain phospholipid liposomal system is presumably because of the influence of its glucose group, resulting in decreased lipophilicity and thus influencing its uptake and orientation within the liposomal membrane. This hypothesis is supported by the much improved ability to inhibit lipid peroxidation demonstrated by the aglycone forms. Bioconversion of phloridzin to 3-hydroxyphloridzin greatly enhanced its ability to inhibit lipid peroxidation. 3-Hydroxylation of phloretin enhanced its ability to inhibit lipid peroxidation by 10-fold, making it of comparable effectiveness to quercetin.

The antioxidant action of oestrogens such as 17β-oestradiol is likely to contribute to their cardioprotective effects. Phloretin, the aglycone form of phloridzin, is oestrogenic (Miksicek 1993) and its 3-hydroxy derivative may have greatly diminished oestrogenic activity that, together with its enhanced antioxidant properties, could enable it to act like the antioxidant antioestrogen/weak oestrogen drug tamoxifen (Ridgway et al. 1996b; Ridway et al. 1997). In the microsomal system, 3-hydroxyphloretin was more effective than 17β-oestradiol. Phloridzin itself was not an effective inhibitor of lipid peroxidation in the systems tested, but phloretin was as effective as 17β-oestradiol in the microsomal system.

Clearly, though, aspects such as the ability of phloridzin to induce glycosuria (the prevention of glucose resorption in the kidney) and the inhibition of sodium-dependent active glucose transport would need to be carefully considered. It seems unlikely, however, that dietary levels would approach the levels required to induce detrimental physiological effects. Moreover, a wide range of dietary flavonoids, in addition to phloridzin have now been shown to be inhibitors of glucose transport (Vedavanam et al. 1999).

Phloridzin extraction from apple tissue by aqueous solvent

Phloridzin is currently available commercially, but at high price. To enable its potential use as a parent compound for food additive production, cheap alternative purification methods have been developed (Ridgway and Tucker 1996a). Some simple effective methods of purification of phloridzin from apple tissues are now considered. The actual methods utilised would depend on a number factors including the value of the derivatives, wage costs and whether on-site or centralised initial extraction was performed.

A particularly effective method, for relatively small-scale laboratory use, or the production of high-value derivatives was found to involve the drying of apple leaves followed by the extraction of phloridzin into ethyl acetate (containing 5% methanol by volume), filtration and crystallisation via the addition of 0.7 volume of chloroform and 0.02 volume of water (Ridgway and Tucker, 1996a). Pure crystals of phloridzin were then obtained by recrystallisation (typically twice) from water (1 g of phloridzin in 100 ml water is a useful working ratio). Freeze-drying aids laboratory procedures, although slow atmospheric drying results in little or no loss of phloridzin yield, providing that the temperature is kept moderate ($<12\,°C$) and exposure to light is low. A typical mid-season (July–early August UK) yield from leaves of the cultivar Bramley's Seedling and a 10:1 ethyl acetate: dry leaf extraction ratio and two water recrystallisations, was 0.9 g phloridzin from 100 g leaves (Ridgway and Tucker 1996a).

For on-site or near-site extraction, crystallisation of phloridzin from hot water ($>80\,°C$) extracts is possible, if tissue disruption is minimal, i.e. whole leaves and not too small twig segments ($>0.25\,mm$) are used (Ridgway and Tucker 1996a). Concentration of filtered extract (by open or reduced pressure evaporation) is then necessary to take the phloridzin concentration below the supersaturation limit of phloridzin ($1\,g/240\,ml$)

and compensate for the impurity impedance of crystallisation. Under these conditions, a 100 g mid-season leaf extract would require concentration to 40 ml from an initial extraction volume of 600 ml to achieve crystallisation, a process that may take over 24 hours to complete. Phloridzin purification following solubilisation into ethyl acetate containing 5% methanol by volume, can then proceed as described above. An alternative hot water extraction scheme can be used to deliberately prevent crystallisation (Ridgway and Tucker 1996a). This is achieved by grinding the tissue in hot water to maximise impurity impedance. Following filtration of the bulk material and concentration down to very low volumes (e.g. 20 ml for a 100 g extract) an almost black phloridzin solution or crude 'apple oil' is produced. Phloridzin can then be extracted by partition against ethyl acetate and purification can proceed (see above), although to achieve full purification, additional organic steps may be required. Typical mid-season leaf yields using initial hot water extractions of Bramley's Seedling, and 10:1 initial water:tissue extraction ratio are 0.8 and 0.6 g respectively for the two hot water extraction methods described above (Ridgway and Tucker 1996a).

The most important determinant of phloridzin content is its period of development and this has been demonstrated by the twig and leaf yields from Bramley's Seedling cultivar (Ridgway and Tucker 1996a). Cultivar, including rootstock type, was found to have much less effect. However, the effect of cultivar can be dramatic in relation to total tissue growth. This gives wide variations in total phloridzin production by given cultivars, and this is illustrated by the results obtained for different rootstocks (Ridgway and Tucker 1996a). Production of phloridzin is best carried out by the coppicing of fast-growing rootstocks, such as M25 or MM106. This is comparable to the production of willow or poplar for biomass, for which mechanical harvesting equipment has been developed. Yields of 250 kg of phloridzin/hectare should then be readily obtained. Development of simple purification methods and the availability of fast-growing rootstock, large scale, and economic production of phloridzin could be carried out. Indeed it is likely that phloridzin production can lead to the development of apple as a major new agrochemical crop (Ridgway and Tucker 1996a), there being much current interest in such crops in the context of maintenance of the rural economy, 'land set aside' and the development of renewable technologies. In the future, it may be possible to further improve yields of phloridzin via the use of genetic engineering (see below). Approaches through biomolecular nutrition and food science will provide novel applications for these apple-derived compounds.

Production of 3-hydroxyphloretin using polyphenol oxidase

The 3-hydroxylation of phloridzin has been achieved using L-ascorbic acid to partially block a polyphenol oxidase reaction. Apple-derived polyphenol oxidase is particularly effective at catalysing this hydroxylation step (Ridgway and Tucker unpublished results). This is a reaction which many forms of polyphenol oxidase such as commercial fungal tyrosinase, do not carry out effectively. Details of methodology for preparation of this polyphenol oxidase are given in Chapter 7.

Polyphenol oxidase has been implicated in enzymatic browning in a number of plant tissues, including potato tubers, bananas, grapes, pears, green olives, kiwi fruit, strawberries, plums and apples (Nicolas *et al*. 1994). Apple is one of the most common fruits in which enzymatic browning, generally considered to be an undesirable reaction because of the unpleasant appearance and development of an off-flavour, is important from the consumer and food industry viewpoint (Nicolas *et al*. 1994). This enzymatic browning occurs when plant tissues are damaged, because polyphenol oxidase then catalyses the oxidation of phenolic compounds to quinones (in the presence of oxygen), which then condense to form darkened pigments: this is an important economic problem. Inhibiting this reaction is feasible by a number of strategies (Walker and Ferrar 1995). For example, by the addition of sulphiting agents: health concerns over sulphites have led to the search for alternatives to these chemical methods. Chelating agents such as EDTA may be used as inhibitors of polyphenol oxidase and are thought to either bind to the active-site copper or reduce the level of copper available for incorporation into the haloenzyme. A copper-binding metallothionein from *Aspergillus niger* has also been reported to inhibit polyphenol oxidase (mushroom tyrosinase) activity (Goetghebeur and Kermasha 1996). Antisense inhibition of polyphenol oxidase gene expression, using constitutive promoters (e.g. CaMV 35S) to express antisense polyphenol oxidase RNA, permits melanin formation to be specifically inhibited in the potato tuber (Bachem *et al*. 1994). The lack of bruising sensitivity achieved in transgenic potatoes suggests that this is a new possibility for the prevention of enzymatic browning in a wide variety of food crops.

In the bioconversion of phloridzin, the L-ascorbic acid acts by blocking the formation of quinones; it achieves this by continually reducing them to the *o*-diphenol form as they are formed. Recovery of the 3-hydroxyphloridzin is by partition against ethyl acetate and precipitation with chloroform, followed by water recrystallisation, i.e. essentially the organic phloridzin purification procedure described above. Conversion of 3-hydroxyphloridzin to the aglycone form results in the production of the highly potent antioxidant 3-hydroxyphloretin. The apple polyphenol oxidase used shows optimum activity at approximately 30% oxygen saturation of water. The design of the reaction vessel used to produce the phloridzin oxidation products took this into account, in addition to its function in replenishing the oxygen consumed in the reaction. With respect to the desirability for relatively slow reaction rates, the optimum design was a 'balanced oxygen type'. This consisted of a stirred tank in which oxygen uptake by stirring was balanced at 30% saturation by the oxygen used as part of the reaction process. In practice (because the reaction rate is determined by oxygen uptake) excess enzyme may be added and the tank stirred at a defined rate. Polyphenol oxidase is inactivated as a result of its own activity with the result that fixing it to a support material would be uneconomic for this procedure, especially considering the loss of activity then incurred (the development of a suitably protective antioxidant support might be useful here).

3-Hydroxyphloridzin could also be manufactured *in situ* in apple juice by the addition of L-ascorbic acid followed by stirring with the pressed apple pulp, which contains endogenous polyphenol oxidase (Ridgway and Tucker 1997c). Increasing the flavonoid levels in apple produced by the use of cider apples because these have a much higher flavonoid content than dessert and culinary types (Lea 1984, 1995), could be another improvement. A potential problem with this strategy, however, is that it may lead to an unacceptably astringent and bitter taste. Flavonoid levels may also be increased by harvesting fruit at an early stage of development: using the apple cultivar Bramley's Seedling leads to an increase in the proportion of phloridzin from 10 to 35% of total phenolic content.

Genetic transformation of plants: prospects for modification of flavonoid antioxidant content

Flavonoids may be beneficial to human health, and the genetic transformation of plants may greatly contribute to this benefit. The prospects for modifying the flavonoid content of plants is discussed (see below), with particular reference to apple as the requisite transformation, and regeneration systems for apple have been developed (James *et al.* 1994). Interestingly, flavonoids may aid the actual process of transformation: a number of chalcones are most effective in *vir* gene induction in *Agrobacterium*-mediated gene transfer (Joubert *et al.* 1995). Apple has been considered as a potential source of phenolics such as phloridzin. Evidently, the economics of phloridzin production would be more favourable if the tissue yields of phloridzin could be improved. Potential schemes for this include the use of regulatory elements, for example, maize R genes, which may coordinately regulate flavonoid biosynthesis, or alternatively the upregulation (expression of non-endogenous forms under a constitutive promoter) of the genes for PAL and CHI, which form potential regulatory points in flavonoid biosynthesis is another possibility.

α-Diphenols undergo greater rates of auto-oxidation than monophenols: the lower levels of stress associated with monophenols may possibly be one of the reasons why apple is able to accumulate large quantities of the monophenolic phloridzin. To increase the accumulation of flavonoids in plants, it may be necessary to upregulate regulatory antioxidant systems such as the glutathione system. This could be carried out by the transformation of plants with bacterial genes such as *gor*, *gsh I* and *II*, coding for glutathione reductase, γ-glutamylcysteine synthetase and glutathione synthetase respectively. In addition, the downregulation of genes coding for L-ascorbic acid oxidase could help to maintain a high antioxidant level. Similarly, the downregulation of the gene(s) coding for polyphenol oxidase, as successfully carried out in potato (Bachem *et al.* 1994), may lead to greater flavonoid stability and accumulation of these health-promoting components. Regulatory elements both *cis* and *trans*, may be affected by redox status (Garcia-Olmedo *et al.* 1994). Increasing antioxidant ability may then be a prerequisite for boosting levels of secondary metabolites, in particular flavonoids. Flavonoids (especially oxidised flavonoids) may interact with cyclin protein. After a critical level of flavonoid accumulation, cell division may cease and cell expansion initiated, thus limiting the flavonoid content of the tissue. A requirement for flavonoid auto-oxidation is possible and this would again emphasise the importance of high cellular antioxidant activity.

In addition to raising levels of flavonoids in plants, there is potential to increase levels of specific flavonoids such as quercetin, which have good antioxidant properties (see above). Quercetin, is known to be present in apple, particularly in the skin of the fruit. This flavonoid is synthesised from dihydroquercetin and represents the end product in this particular biosynthetic pathway. Levels of quercetin may potentially be increased by antisensing the gene for dihydroflavonol reductase. This enzyme competes for available dihydroquercetin with the enzyme responsible for quercetin production and down-regulation of dihydroflavonol reductase may thus lead to a diversion of metabolites into quercetin: the gene for dihydro-flavonol reductase appears to be highly conserved and exists as a single copy.

Flavonoids as 'bioparadox modulators': relevance to apoptosis

Paradoxically antioxidant action may not always be beneficial: apoptosis (programmed cell death: see Chapter 5) can be initiated by oxidative damage, and when cancer cells are exhibiting oxidative stress, but are unable to achieve an apoptotic event (Toyokuni *et al.* 1995), insufficient pro-oxidant activity may be available for self-destruction of the tumour. Antioxidant potential could thus hinder important destructive cellular processes (indeed it may have contributed to the reported increase in lung cancer among smokers given antioxidant β-carotene supplements). Flavonoids, however, may overcome this apparent paradox as they both exhibit antioxidant properties (and under some conditions pro-oxidant properties) and can limit cell division (Ridgway and Tucker 1997b). A possible mechanism for this is 'phenolisation' of cyclin proteins, which is in contrast to the usual regulatory phosphorylation mechanism. This process could occur by π–π binding or covalently *via* the formation of oxidised flavonoid-derived semi-quinones (Ridgway and Tucker 1997b). One possibility is the 'phenolisation' of exposed tyr-15 on p34[cdc2], which would lead to arrest at the G2-M phase of mitosis; phosphorylation at this point is a checkpoint control (Jacobs 1995). G2-M phase arrest is observed in the breast cancer cell line MDA–MB468 when exposed to quercetin (Avila 1994). The isoflavone genistein, in addition, has the potential to interfere directly in the process of phosphorylation as it can inhibit tyrosine kinase activity (Akiyama *et al*. 1987).

Antioxidant activity can be regarded as susceptibility to oxidation by the compound itself. Readily oxidised flavonoids such as 3-hydroxyphloretin are therefore more likely to interact with target sites, such as exposed tyrosine residues, than non-antioxidant compounds. It may be possible to enhance such effects by the use of polyphenol oxidase active fragments linked to tumour-specific antibodies (Ridgway and Tucker 1997b). In high concentrations, it is likely that oxidised phenolics will produce a potentially useful direct toxic effect, and redox cycling to generate toxic free radicals is the basis of the current use of quinonic pharmaceuticals in chemotherapy (Powis 1989). The success of this strategy would again appear to depend on the differential response of cells: cancer cells are unable to cope with high oxidative stress. The redox potential of many existing (and new) biochemicals will thus emerge as a design feature of antioxidants (and pro-oxidants).

Biotransformation of flavonoids: the future in antioxidant applications

Applications of flavonoid oxidation have been discussed, from *in vitro* bioconversions (mediated by polyphenol oxidase) for the production of new food additives, to bioconversions in plant material during food processing, and even transformation in the body, both as part of a redox couple to prevent oxidative damage and oxidation to help destroy tumours. The effectiveness of phloridzin derivatives as inhibitors of lipid peroxidation suggests their use as functional foods (to protect the health of the consumer against the oxidative damage implicated in many disease states). The antioxidant activity of 3-hydroxyphloretin (Ridgway *et al* 1997a,b), clearly shows its potential for use as a therapeutic antioxidant. Its use is likely to prove most effective in combination with other types of antioxidant, not just aqueous-phase types such as L-ascorbic acid (vitamin C with added bioflavonoids is already to be found on sale as a food supplement), but perhaps also with membrane fluidity modifiers such as tamoxifen (Wiseman 1994a).

The weak oestrogenic properties of phloridzin and its derivatives suggest that they could act in a similar way to soy-derived isoflavones to block the mitogenic effects of oestrogen on breast cells and thus act as dietary chemopreventative agents against breast cancer (Ridgway *et al*. 1997a,b).

Oestrogenic properties of the 3-hydroxy derivatives of phloridzin have yet to be reported. It is likely, however, that the 3-hydroxy derivatives will have lower oestrogenic activity (Miksicek 1993, 1994, 1995). Other useful direct bioconversions of phloretin include methylation, for example, by the use of naringenin 7-o-methyltransferase (Rakwal et al. 1996). This would perhaps relate to improving oestrogen antagonist and hence anticancer activity. These novel flavonoids could provide the standard technological anti-oxidant function of preventing rancidity in fatty foods.

Dietary modulation of membrane fluidity and of enzyme and receptor activity

Changes in membrane receptor and enzyme accessibility may arise as a result of modulation of membrane fluidity (Viret et al. 1990). Membrane lipid peroxidation results in loss of polyunsaturated fatty acids, decreased membrane fluidity and severe structural changes resulting in loss of enzyme and receptor activity (van Ginkel and Sevanian 1994) and direct free radical damage to membrane proteins may also occur as a result of lipid peroxidation (Oliver et al. 1987; Dean et al. 1993; Stadtman 1993) leading to their inactivation. Moreover, dietary polyenylphosphatidylcho-line (a choline glycerophospholipid containing up to 80% of total fatty acids as linoleic acid) can induce the activity of the enzyme delta-6-desaturase in rat liver under conditions of oxidative stress, and the action of this enzyme is protective against the polyunsaturated fatty acid destroying effects of lipid peroxidation (Biagi et al. 1993). It may appear somewhat contradictory that some membrane antioxidants appear to inhibit lipid peroxidation through a mechanism that appears to involve, at least in part, decreased membrane fluidity. However, this subtle modulation of membrane fluidity by these dietary components may have beneficial influences on membrane receptor and enzyme activity in disease states such as cancer.

The activity of the membrane-bound enzyme 5'-nucleosidase, was found to be significantly higher when the rats were fed an olive oil-based diet compared to when they were fed a blackcurrant oil-based diet (Barzanti et al. 1995). Olive oil is rich in the monounsaturated oleic acid compared to blackcurrent oil, which is rich in linoleic and 18C-polyunsaturated fatty acids but deficient in oleic acid. The fatty acid composition of the olive oil-based diet may predispose directly towards lower membrane fluidity and also less lipid peroxidation than the blackcurrant oil, and it is possible that this influences membrane enzyme activity (Barzanti et al. 1995). Dietary components that alter membrane fluidity either through direct membrane interaction or indirectly through alteration of susceptibility to lipid peroxidation may influence cell signalling and this is particularly important in disease states such as cancer (see below). The activity of the enzyme adenylate cyclase, which catalyses the formation of the second messenger cyclic AMP in response to hormones and growth regulators binding to cell surface receptors, is very sensitive to membrane fluidity (Houslay 1985). Increasing the cholesterol content of isolated plasma membranes caused a progressive decline in adenylate cyclase activity probably because of restricted mobility of the components of the enzyme complex (Houslay 1985). Furthermore, the membrane phospholipid environment has been reported to be important in receptor interaction with the adenylate cyclase complex in intact cell membranes (Jansson *et al.* 1993).

Protein kinase C is a calcium- and phospholipid-dependent enzyme that mediates the transmembrane signalling of a wide range of extracellular stimuli (including growth factors, hormones and other biologically active substances) resulting in the phosphorylation of various cellular proteins believed to be involved in proliferative control (see below). Inhibition of protein kinase C may also result from decreased membrane fluidity. Indeed, the likely importance of protein kinase C being able to undergo conformational change upon activation has been hypothesised (Gschwent *et al.* 1991; Burns and Bell 1992). It is of related interest that a protein kinase C-dependent inhibition of lipid peroxidation appears to exist that is normally inhibited by the membrane antioxidant vitamin E, and is switched on only upon consumption of vitamin E in free radical reactions (Kagan *et al.* 1992). Other pathways that may be influenced by diet-induced changes in membrane fluidity include the sphingomyelin pathway (Kolesnick and Golde 1994). This pathway is initiated by hydrolysis of plasma membrane sphingomyelin by sphingomyelinase to the second messenger ceramide. The ceramide-activated serine/threonine protein kinase transduces a cytokine signal, partly through mitogen-activated protein (MAP) kinase and transcription factors such as NF-κB, and the sphingomyelin pathway is thought to be utilised by both tumour necrosis factor α and interleukin-1β to effect signal transduction by their receptors (Kolesnick and Golde 1994).

The insulin receptor is located in the outer membrane domain and possesses tyrosine kinase activity: the crystal structure of this domain has been determined (Hubbard *et al.* 1994). The activity of this receptor has been suggested to be related to the fluidity in that part of the membrane (Berlin *et al.* 1994; McCallum and Epand 1995). Membrane fluidity-mediated changes in conformation of the insulin receptor may have caused the observed changes in insulin binding to erythrocyte ghosts, in humans and in monkeys, induced by modifications in dietary fat (Bhathena *et al.* 1996; Barnard *et al.* 1990). Increased consumption of polyunsaturated fatty acids resulted in higher fluidity of human erythrocyte membranes and higher insulin binding (Berlin *et al.* 1994). Furthermore, dietary substitution of saturated fatty acids with *cis-* or *trans-* monounsaturated fatty acids has been shown to have no influence on insulin binding (Berlin *et al.* 1994).

Modulation of membrane enzyme activity by lipid peroxidation has been studied in reconstituted membranes consisting of intestinal alkaline phosphatase and phosphatidylcholine or dipalmitoylphosphatidylcholine: loss of enzyme activity was observed on lipid peroxidation (Ohyashiki *et al.* 1994a). Lipid peroxidation and decreased membrane fluidity have also been shown to decrease the reactivity of the thiol groups of membrane proteins in a porcine intestinal brush border membrane system with a fluorescent probe (Ohyashiki *et al.* 1994b).

Genistein is a specific inhibitor of tyrosine kinases (Akiyama *et al.* 1987; Holting *et al.* 1995) and may additionally influence the activity of other membrane enzymes through modulation of membrane fluidity (see above). Moreover, hydrogen peroxide has been shown to enhance the tyrosine kinase-mediated phosphorylation of the epidermal growth factor receptor in both intact cells and a solubilized membrane fraction (Gamou and Shimizu 1995) again indicating the role of oxidative stress in cell signalling processes through the cell membrane that may be involved in cancer (see below).

Dietary modulation of membranes in disease treatment and prevention?

Membrane modulation in anticancer action and reversal of multi-drug resistance

The ability of dietary components such as the isoflavonoid phytoestrogens and vitamin D to exert their membrane antioxidant action at least partly through decreased membrane fluidity could inhibit the activity of membrane enzymes such as protein kinase C (Wiseman, 1994a) and adenylate cyclase (Houslay 1985). In relation to inhibition of cancer cell growth, inhibition of adenylate cyclase would decrease cellular cAMP levels, found to inhibit the growth of some but not all cancer cells, and this is an area that has attracted considerable controversy (Dumont *et al.* 1989). Protein kinase C activity has also been suggested to be inhibited by modulators of membrane fluidity (see above) and this would have important consequences for cancer cell growth (Wiseman 1994a). The possible role of dietary flavonoids and soya phytoestrogens in cancer prevention is discussed in Chapter 6. Vitamin D intakes (in association with calcium levels) may be important in protection against colon cancer (Lipkin and Boone 1991) and synthetic vitamin D analogues appear to inhibit the growth of breast cancer cells both *in vitro* and *in vivo* (Colston *et al.* 1992). Although vitamin E has been suggested to be protective against cancer, a large prospective study showed that high dietary intakes of vitamin E did not appear to protect women against breast cancer (Hunter *et al.* 1993).

Membrane modulation may play an important role in the ability of drugs and possibly dietary factors to reverse multidrug resistance (MDR). MDR is the phenomenom in cancer chemotherapy whereby tumours that initially show sensitivity to a particular treatment become resistant to a wide range of anticancer drugs (Pasman and Schouten, 1993). A MDR phenotype can result from the overproduction in resistant cells of a transmembrane P-170 glycoprotein (other types including non-*p*-glycoprotein-mediated resistance also exist). This membrane protein appears to contribute to MDR through an ATP-dependent drug efflux mechanism, which pumps cytotoxic drugs (e.g. doxorubicin and vinblastine) out of the cell (thus preventing their accumulation to an effective cytotoxic concentration). Pharmacological agents have been identified that antagonise the action of the P-170 glycoprotein including tamoxifen (Leonessa *et al.* 1994). Decreased membrane fluidity in the presence of tamoxifen could directly inhibit function of the P-170 glycoprotein, possibly by altering its conformation (Wiseman, 1994a; Leonessa *et al.* 1994). In addition, tamoxifen has been reported to decrease drug efflux from the liposomes (Kayyali *et al.* 1994). This membrane model bears some resemblance to a cell thus inhibition of a similar basal mechanism for drug removal in cancer cells may contribute to the anti-MDR action of tamoxifen (Kayyali *et al.* 1994). Dietary components that decrease membrane fluidity may act in a similar way and thus be beneficial against MDR. Indeed, genistein has been reported to reverse non-*p*-glycoprotein mediated MDR (Takeda *et al.* 1994).

Increased susceptibility of cancer cells to oxidative membrane damage through dietary mediation

Cancer cells obtain all of their n-6 and n-3 polyunsaturated fatty acids from the circulation of their host and thus dietary fat intake will influence availability (Burns and Spector 1994). It may be possible to change the membrane composition (particularly the polyunsaturated fatty acid content) and thus both the membrane fluidity and susceptibility to oxidative damage of cancer cells by altering the dietary fat intake of the host (Burns and Spector 1994). Membrane fluidity would be increased by enriching the membranes with polyunsaturated fatty acids and this would influence the activity of membrane-bound enzymes, receptors and transport mechanisms. It has been predicted that cancer cells with membranes enriched with these fatty acids would be more susceptible to oxidative membrane damage. Indeed, L1210 murine leukaemia cells enriched with 22:6 fatty acids produced an intense radical-adduct spectrum when subjected to oxidative stress (Wagner *et al.* 1993) and produced increased amounts of TBARS (Wagner *et al.* 1992).

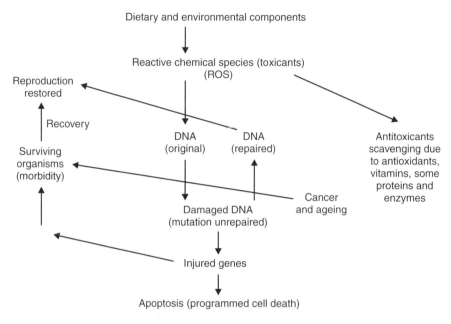

Figure 1.2. Life/death cycle caused by biomolecular injury

The n-3 *cis*-unsaturated fatty acids docoshexanoic acid and eicosapentae-noic acid and the n-6 unsaturated fatty acid gamma-linolenic acid are cytotoxic to both vincristine-sensitive (KB-3-1) and resistant (KB-ChR-8-5) human cervical cancer cells in culture and have all been shown to enhance the formation of lipid peroxides, hydrogen peroxide and superoxide radicals in KB-3-1 cells. A desirable increase in oxidative membrane damage in these cancer cells could be mediated by dietary fatty acid intake (Burns and Spector 1994).

Oxidative membrane damage in DNA damage leading to the development of cancer? (see Chapter 2)

The cumulative risk of cancer increases with approximately the fourth power of age, and in humans approximately 30% have cancer by age 85 (Ames 1989). Cancer can therefore be considered to be a degenerative disease of old age and it has been proposed that this is related to the effects of continuous damage over a lifespan by ROS and RNS (Ames 1989). ROS/RNS can cause DNA base changes, strand breaks, damage to tumour suppressor genes and enhanced expression of proto-oncogenes (Cerutti 1994; Wiseman and Halliwell 1996; Poulsen et al. 1998; Loft and Poulsen 1998). However, the development of human cancer depends on many other

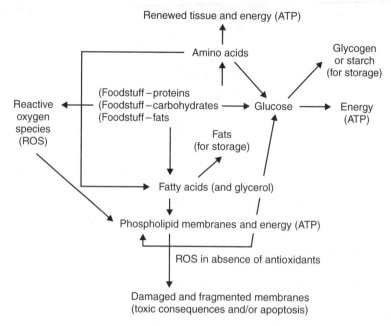

Figure 1.3. Dietary choices related to biomolecular-nutritional outcomes

factors, including the extent of DNA damage, antioxidant defences, DNA repair systems, efficiency of removal of oxidized nucleosides, before they are incorporated into DNA and the cytotoxic effects of ROS in large amounts as well as their growth-promoting effects in small amounts (Cerutti 1994; Wiseman and Halliwell 1996).

Plant polyphenols have a protective effect against radiation-induced chromatid breaks in cultured human cells: green and black tea extracts their polyphenols and curcumin were tested and only the green tea polyphenol (-)-epigallocatechin gallate had no effect (Parshad et al. 1998). This protective effect is likely to arise from their antioxidant action, particularly the scavenging of hydroxyl radicals. Quercetin can prevent DNA single strand breakage (strand scission) in addition to cytotoxicity, suggesting that it is acting predominantly as an iron chelator in this system as free radical scavenging antioxidants prevent only cytotoxicity (Sestili et al. 1998). Recently, however, we failed to find a substantial effect of dietary quercetin (consumed in onions and black tea for two weeks) on oxidative DNA base damage (measured by GC-MS) in healthy human subjects (Beatty et al. 2000).

ROS/RNS-induced mutations could result not only from direct DNA damage but also indirectly as a consequence of oxidative damage to membranes (Wiseman and Halliwell 1996). Attack on membrane lipids can initiate the process of lipid peroxidation and the lipid peroxides formed as a result of this oxidative membrane damage can subsequently decompose to mutagenic carbonyl products (Cheeseman 1993). In an investigation of a baby hamster kidney cell line (BHK-21/C13) and its polyoma virus transformed malignant counterpart (BHK-21/PyYcells) the level of lipid peroxidation was higher in transformed cells than in non-transformed cells, suggesting that the level of lipid peroxidation is increased in the malignant state (Diplock *et al.* 1994; Goldring *et al.* 1993). This is in contrast to earlier work that claimed that susceptibility to lipid peroxidation is decreased in malignant hepatoma cells (e.g. Novikoff and Yoshida ascites hepatoma cells) (Cheeseman 1993). These observations suggest that cell transformation alters cell responsiveness to oxidative stress. Lipid peroxidation has been suggested to have a role in human breast cancer risk: urinary excretion of the mutagen malondialdehyde has been shown to be approximately double in women with mammographic displasia (high risk) than in women without these changes (Boyd and McGuire 1991).

Inflammation can accelerate the development of cancer, including that caused by viral, bacterial and parasitic infections (Rosin *et al.* 1994). In colon cancer, predisposing sources of chronic inflammation include ulcerative colitis and infection with the parasite *Schistosoma japonicum* (Rosin *et al.* 1994). There is considerable evidence that ROS/RNS are involved in the link between chronic inflammation and malignant progression to cancer (Rosin *et al.* 1994; Wiseman and Halliwell 1996). A convincing relationship has been demonstrated between the capacity of tumour promoters to stimulate inflammatory cells to release ROS/RNS and their capacity to promote tumours. Tumour promotion in animal models can be inhibited by antioxidants capable of inhibiting the phagocyte oxidative burst (Wiseman and Halliwell 1996). This evidence suggests that dietary factors that act as antioxidants to decrease membrane lipid peroxidation may contribute to a protective effect against the development of cancer, including malignant progression to cancer arising from chronic inflammation.

Oxidative damage to membranes and neurodegeneration, neurotoxicity and ageing

The free radical hypothesis of ageing proposes that the harmful actions of ROS are responsible for the functional deterioration found with ageing. Furthermore, as cellular membranes (particularly mitochondrial) are prone to the greatest damage from ROS-mediated damage, it has been suggested that oxidative damage to membrane lipids is intimately involved in the ageing process (Rikans and Hornbrook 1997; Knight 1998). Although lipid peroxidation and decreased antioxidant protection are often found to be associated with ageing, this is not always the case. Indeed, age-dependent changes in lipid peroxidation and antioxidant protection appear to be sex, species and tissue specific (Rikans and

Hornbrook 1997). Future studies are clearly needed that use better measures of *in vivo* lipid peroxidation and clearly identify the likely critical targets of lipid peroxidation.

Oxidative damage to mitochondria, including damage to the mitochondrial membrane has been implicated in neurodegenerative disorders such as Parkinson's disease and ageing (Shigenaga *et al.* 1994; Jenner 1994). Intermediate radicals formed during lipid peroxidation can also attack DNA (Cheeseman 1993) and have been suggested to damage mitochondrial DNA, which is close to the inner mitochondrial membrane (Hruszkewycz 1992). This injury mediated by oxidative membrane damage could contribute to the deletions and mutations in mitochondrial DNA that accumulate with age at a higher rate than in nuclear DNA. Damage to mitochondrial DNA could play a role in neurodegenerative diseases: mitochondrial deletions and increased mitochondrial oxidative DNA damage have been reported in Alzheimer's disease (Mecocci *et al.* 1995). Moreover, in the mouse, mitochondrial oxidant production and associated damage and ageing are associated with enhanced food consumption (Sohal *et al.* 1994).

Neural membranes are highly susceptible to oxidative damage, and vitamin E is the most important antioxidant in neural membranes: vitamin E depletion may result in oxidative injury and dietary vitamin E supplementation can enrich brain membranes and provide a higher level of protection against lipid peroxidation (Kagan *et al.* 1992). Lipid peroxidation in brain membranes arising from exposure to lead in rats (50 mg/kg body weight intragastrically for eight weeks) was positively correlated with inhibition of the activity of the membrane-bound enzyme acetylcholinesterase (Sandhir *et al.* 1994). This suggests that dietary lead may exert its neurotoxic effects via oxidative membrane damage and subsequent loss of membrane function. Ageing is associated with a progressive decrease in selenium status (measured as serum selenium and glutathione peroxidase activity) and in the ratio of plasma and erythrocyte membrane polyunsaturated to saturated fatty acids (Olivieri *et al.* 1994). This supports the idea that oxidative damage to membranes both accumulates with age and is implicated in the ageing process. Cataract formation is an important consequence of accumulated free radical damage that manifests with age. A study of the causes of membrane damage in UV-irradiated eye lenses showed increased formation of lipid peroxidation products, suggesting that damage to the lens membrane plays an important role in the overall damage effect (Hightower *et al.* 1994). Dietary membrane antioxidants such as vitamin E may offer protection against this age-related disorder.

Lipoprotein particle injury in atherogenesis

Dietary components may protect human low-density lipoproteins (LDL) against oxidative modification. This may be of importance because

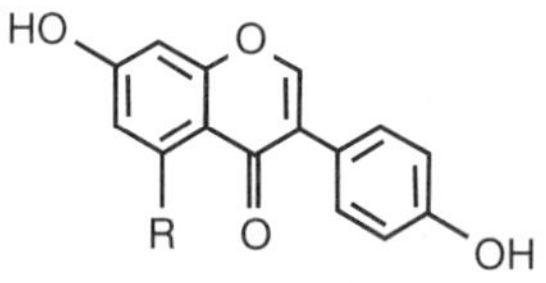

Daidzein R = H
Genistein R = OH

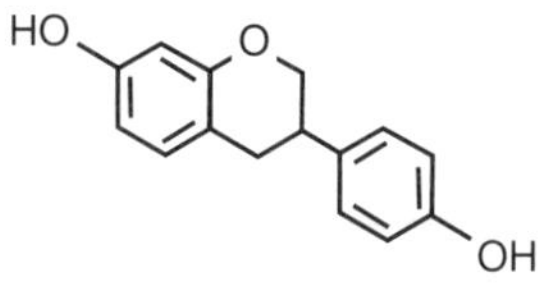

Equol

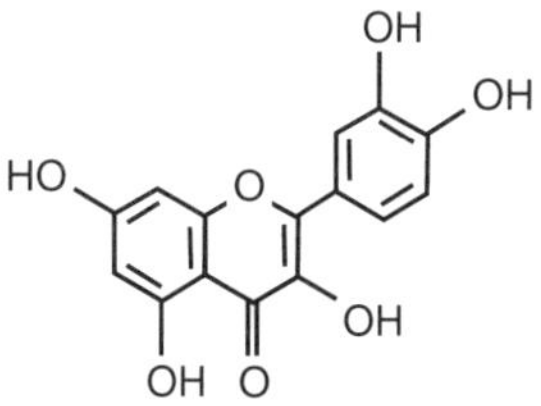

Quercetin

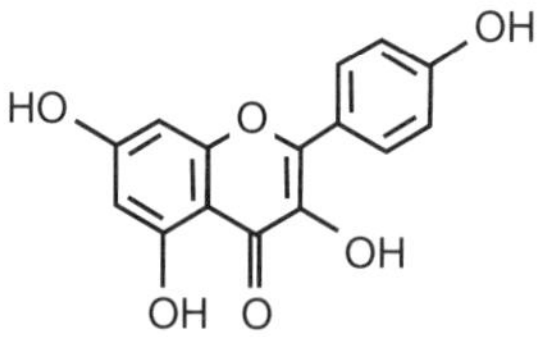

Kaempferol

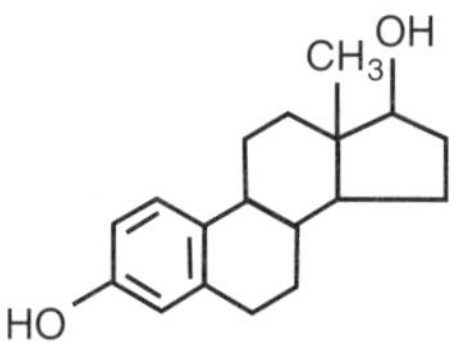

17β-Oestradiol

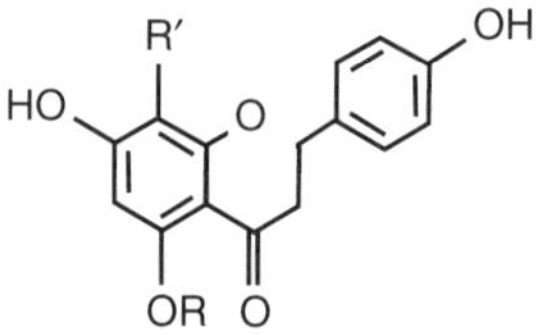

Phlorizin R = β-D-glucose R¹=H
Phloretin R=R¹=H

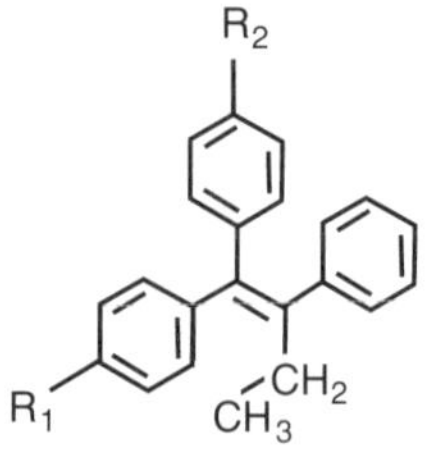

Tamoxifen R₁=H R₂=OCH₂CH₂N(CH₃)₂
4-Hydroxytamoxifen R₁=OH R₂=OCH₂CH₂N(CH₃)₂

Figure 1.4. Examples of membrane antioxidant structures

oxidative damage to LDL (particularly to the apoprotein B molecule) is an important stage in the development of atherosclerosis: it is a prerequisite for macrophage uptake and cellular accumulation of cholesterol leading to the formation of the atheromal fatty streak (Witzum 1994). Lipid peroxidation starts in the polyunsaturated fatty acids of the phopholipids on the surface of LDL and then propagate to core lipids resulting in modification of the cholesterol, phospholipids and the apolipoprotein B molecule, and also to the polyunsaturated fatty acids (Witzum 1994, Halliwell 1995b). Furthermore, linoleic acid peroxidation has been shown to be the dominant lipid peroxidation process in LDL (Spiteller 1998). Oxidised LDL is highly atherogenic because it stimulates macrophage cholesterol accumulation and subsequent foam cell formation. In addition, it is cytotoxic to the cell of the arterial wall and can stimulate thrombotic and inflammatory processes. Macrophage-mediated oxidation of LDL occurs in early atherosclerosis and depends on the oxidative state of not only the LDL but also of the macrophages. Factors that decrease the LDL oxidative state include the presence of vitamin E, β-carotene and polyphenolic flavonoids and the latter can also decrease the macrophage oxidative state and consequently cell-mediated LDL oxidation (Aviram and Fuhrman 1998).

LDL is usually either isolated following a dietary study or dietary components are added to isolated LDL *in vitro*. In most of these studies on the action of dietary components on oxidative damage to LDL, human LDL is stimulated to undergo lipid peroxidation by the addition of Cu(II) ions: an experimental system that is relevant to events occurring within the atherosclerotic lesion (Smith *et al.* 1992; Halliwell 1995b). These events appear to involve pre-existing lipid hydroperoxides (Thomas *et al.* 1994; Halliwell 1995b). Copper(II) ions have been shown to stimulate LDL oxidation by different mechanisms, depending on the concentration used: a marked change in the kinetics of LDL oxidation was observed at submicromolar Cu(II) concentrations (Ziouzenkova *et al.* 1998).

Lipid peroxidation in membranes and in lipoprotein particles, including LDL, proceeds through initiation, propagation and termination phases characteristic of free radical damage. These are expressed by a lag phase during which not much oxidation occurs, followed by a rapid increase in autocatalysis by chain-propagating intermediates and concluding with a decrease in the rate of oxidation. The lag phase, which can be measured (e.g. for LDL oxidation by continuous measurement of the formation of conjugated dienes) can be increased by chain-breaking antioxidants, which scavenge the initiation reaction or intercept chain-propagating species, thus the lag phase in lipid peroxidation processes reflects the antioxidant status of membranes and lipoproteins and their oxidation resistance (Cadenas and Sies 1998). Furthermore, measurement of the LDL lag time is often used to test the efficacy of a wide range of antioxidants for ability to protect LDL against oxidation (Parthasarathy *et al.* 1998). The antioxidant content of human LDL is clearly of great importance in protecting against oxidation and endogenous LDL antioxidants include α-tocopherol, γ-tocopherol, β-carotene and lycopene and coenzyme-Q10 (ubiquinol-10) (Tertov *et al.* 1998). Vitamin E is an important antioxidant against LDL oxidative modification, although the full mechanisms of vitamin E action in LDL have yet to be elucidated total antioxidant potency is likely to be determined by the reactivities and concentrations of substrates, radical and antioxidant, and thus the relative importance of the many competing reactions (Noguchi and Niki 1998).

Nitric oxide can have both pro-oxidant and antioxidant effects on LDL. Although nitric oxide does not appear to react directly with LDL, in the presence of oxygen or superoxide (through peroxynitrite formation) it may cause oxidation of LDL lipid, protein and antioxidants, but when LDL oxidation is initiated by copper ions or azo-initiators, nitric oxide is a potent oxidation inhibitor (Hogg and Kalyanaraman 1998). Furthermore, when released slowly from a donor compound LDL has a potent inhibitory effect on the oxidative modification of LDL (Goss *et al.* 1998).

The oxidative hypothesis of atherosclerosis and the effectiveness of dietary antioxidants appears to be supported by evidence from clinical trials (Nyyssonen *et al.* 1994; Hoffman and Garewell 1995), although further studies with biomarkers that accurately reflect the extent to which antioxidant intervention actually inhibits atherosclerosis are clearly still needed (Schwenke 1998). Furthermore, therapeutic perspectives in relation to LDL oxidation need to be considered because some interventions aimed at decreasing LDL oxidative susceptibility have not been shown to attenuate atherogenesis (Heller *et al.* 1998). Furthermore, in animal studies, extent of protection of plasma LDL by antioxidants is not a predictor of the extent of their antiatherogenic effects (Fruebis *et al.* 1997).

Dietary modulation of LDL resistance to oxidation

Epidemiological evidence from the Dutch Zutphen elderly study, suggests that flavonoid consumption is associated with a lower risk of coronary heart disease (Hertog *et al.* 1993; Hertog and Hollman 1996). However, this has not been confirmed by subsequent epidemiological studies: the US male health professional non-fatal myocardial infarction study (Rimm *et al.* 1996) and the Caerphilly (Welsh males) ischaemic heart disease study (Hertog *et al.* 1997). Nevertheless, flavonoids in red wine protect LDL against oxidative damage (Furhrman *et al.* 1995, Whitehead *et al.* 1995) and the antioxidant properties of flavonoids may contribute to the reduced risk of coronary heart disease in wine drinkers (despite their high-fat diet and smoking habits): this is the so-called French paradox (Renaud and De Lorgeril 1992).

Flavonoids (morin, quercetin, fisetin, galangin and chrysin) have been reported to inhibit the oxidative modification of low-density lipoproteins by macrophages (de Whalley *et al.* 1990). Most *in vitro* studies show quercetin to be the most effective flavonoid inhibitor of copper ion-mediated LDL lipid peroxidation (O'Reilly *et al.* 2000). Phenolic acids, for example, caffeic acid, chlorogenic acid, ellagic acid and protocatechuic acid all protect isolated LDL against oxidative damage (Laranjinha *et al.* 1994). However, consumption of quercetin-rich foods does not appear to protect against LDL oxidation *ex vivo* (O'Reilly *et al.* 2000).

trans-Resveratrol (a stilbene and phytoestrogen) is another phenolic compound present in some red wines, and is thought to help protect against atherosclerosis by protecting LDL against oxidative damage. The ability of *trans*-resveratrol to protect LDL against oxidative damage has been found to be mostly because of its capacity to chelate copper, although it also scavenges free radicals. *trans*-Resveratrol inhibits the formation of TBARS in phospholipid liposomes, and together with the lipophilic nature of the interaction of *trans*-resveratrol with plasma lipoproteins this suggests that it may be effective at a number of different sites, i.e. in the lipid and protein moieties of LDL and in their aqueous environment (Belguendouz *et al.* 1998).

(+)-Catechin (Mangiapane *et al.* 1992) and the flavonoid myrigalone B (in combination with ascorbic acid) (Mathiesen *et al.* 1996) also protect LDL against oxidative damage. Phloridzin and its derivatives were able to protect LDL against oxidative damage and for some of these compounds this effect was greater than in model membrane systems. The isoflavones genistein and daidzein, and in particular the isoflavan equol, have been reported to display antioxidant properties *in vitro*, predominantly against oxidative damage to membrane lipids (see above) and lipoprotein particles (Arora *et al.* 1998). Protection of LDL against oxidative damage has been demonstrated for the oestrogen 17β-oestradiol both *in vitro* (Huber *et al.* 1990; Rifici and Khachadurian 1992; Wiseman *et al.* 1993b) and *in vivo* (Sack *et al.* 1994) and for tamoxifen *in vitro* (Wiseman *et al.* 1993b). Genistein, daidein, equol, quercetin, 17β-oestradiol, tamoxifen and 4-hydroxytamox-ifen have all additionally been shown to protect LDL against the glucose-induced lipid peroxidation implicated in the increased prevalence of atherosclerosis in diabetic patients (Vedavanam *et al.* 1999). The overall order of potency for the compounds tested was quercetin=equol> kaempferol>17β-oestradiol>4-hydroxytamoxifen>genistein>tamoxifen= daidzein (Vedavanam *et al.* 1999).

The Mediterranean diet, which is rich in fruit, vegetables, grains and vegetable oil (in particular olive oil) is associated with a lower incidence of coronary heart disease and this may be due at least in part to inhibition of LDL oxidation by the natural antioxidant action of various dietary components. A study comparing antioxidant status and susceptibility of plasma lipoproteins to undergo lipid peroxidation in healthy young persons from Naples, Italy, compared to Bristol, UK, showed that the Naples group consumed more fresh tomatoes, more fat as monounsaturates (from olive oil) and had higher plasma levels of the lipid antioxidants vitamin E and β-carotene, whereas intakes of vitamin C, total uncooked fruit and vegetables, plasma vitamin A, serum selenium and copper levels were similar (Parfitt et al. 1994). All indices of LDL oxidation were significantly lower in the Naples group (Parfitt et al. 1994). Another study of LDL isolated from healthy individuals showed that the antioxidant content (vitamin E and β-carotene), fatty acid composition and intrinsic phospho- lipase A2-like activity had little influence on LDL susceptibility to Cu(II) induced oxidation (Croft et al. 1995). These components did, however, significantly influence both the rate and extent of LDL oxidation with increased vitamin E, linoleic acid content and phospholipase activity associated with faster and more extensive oxidation (Croft et al. 1995). Oleic acid content was negatively correlated with the rate of LDL oxidation (Croft et al. 1995). The apparent pro-oxidant effect observed has interesting implications for the generally accepted protective effects of vitamin E on atherogenesis (Croft et al. 1995).

LDL from obese girls in Japan was found to contain more polyunsatu- rated fatty acids compared to normal controls and thus a higher peroxidizability index (estimates the susceptibility of lipids to oxidative stress) than normal controls, LDL α-tocopherol and β-carotene concentra- tions were also lower than in normal controls, indicating an increased susceptibility of LDL to oxidative stress in obese girls that may increase their risk of atherosclerosis later in life (Kuno et al. 1998).

Increased consumption of green tea (especially more than 10 cups per day) in 1371 Japanese men aged over 40 years increased the proportion of the protective high-density lipoprotein (HDL: more resistant to oxidation than LDL) and a decreased proportion of LDL (Imai and Nakachi 1995) and this may be another effect of the flavonoids and other polyphenols in tea that exert a strongly protective effect against oxidative damage to LDL (see above). In the Penn State Young Women's Health Study, lower ratios of total serum cholesterol to HDL cholesterol, lower weight, lower body mass index and higher fruit, carbohydrate and fibre intakes were found in the first quintile by percentage body fat compared to the fifth quintile (Lloyd et al. 1998). There is interest in the relationship between coffee consumption and adverse alterations in lipoprotein profiles in humans; however, a study of 541 pre-menopausal women whose mean coffee consumption was 3.35 cups of coffee per day showed no noticeable effect (Carson et al. 1994). Oestrogen has been reported to decrease LDL levels and increase HDL levels (Haffner and Valdez 1995) and dietary isoflavones may have similar beneficial effects on lipid profiles (Dean et al. 1998).

Future prospects for antioxidant protection against membrane lipid and lipoprotein injury and for dietary modulation of membrane function in disease prevention

The uptake of drug molecules into lipid membranes is likely to alter the physical properties and function of the membrane, in particular the main gel to liquid crystalline phase transition can be drastically changed and this may lead to changes in the conformation and thus activity of embedded membrane receptor proteins (Seydel *et al.* 1994). Furthermore, specific drug–lipid interactions may lead to drug accumulation in membranes and thus to much larger concentrations at the active site than in the surrounding aqueous phase: in the lipid environment this may lead to changes in the preferred conformation of the drug molecules themselves (Seydel *et al.* 1994).

Dietary components, particularly those with similar properties to known drugs, for example, isoflavone phytoestrogens, which have some similar properties to tamoxifen, may be able to modulate membrane function in this way and this is an area certainly worthy of further study. Major areas of potential for dietary modulation of lipoproteins to achieve protection against disease (atherosclerosis) are firstly in altering lipoprotein profiles to decrease LDL levels and elevate HDL levels, and secondly in altering LDL composition to increase levels of protectants against oxidative damage. Dietary components with both antioxidant and oestrogenic (favourable alteration of lipoprotein profiles likely) properties such as the isoflavone phytoestrogens (Wiseman *et al.* 1998, 2000; Dean *et al.* 1998) seem promising.

Intake of fresh fruit and vegetables, which are the main sources of the dietary antioxidants that may protect against the oxidative damage to membranes implicated in disease, appears to be inversely correlated with particular cancers (Block *et al.* 1992; Wiseman and Halliwell 1996). The optimum level of dietary antioxidants has been suggested to be achievable by the intake of at least three servings of vegetables and two of fruit per day. Furthermore, increased consumption of phytoestrogen containing soy products may also prove to be beneficial in protecting membranes against oxidative damage.

The cell membrane as a target for dietary modulation in disease prevention displays much potential for future investigations and progress. Understanding of the ways in which dietary factors can influence membrane function will be greatly aided by determinations of structure–function relationships and of dietary component–phospholipid and dietary component–protein interactions, perhaps facilitated by the use of computer biomolecular modelling.

References

Akiyama, T., Ishida, J., Nakagawa, S., Ogawata, H., Watanabe, S.-I., Itoh, N., Shibuya M. and Fukami, Y. (1987) Genistein, a specific inhibitor of tyrosine specific protein kinases. *J. Biol. Chem.*, **262**, 5592–5595.

Allard, J.P., Royall, D., Kurian, R., Muggli, R. and Jeejeebhoy, K.N. (1994) Effects of β-carotene supplementation on lipid peroxidation in humans. *Am. J. Clin Nutr.*, **59**, 884–890.

Ames, B.N. (1989) Endogenous oxidative DNA damage, aging and cancer. *Free Radical Res. Communs.*, **7**, 121–128.

Anderson, J.J.B. and Toverud, S.U. (1994) Diet and vitamin D: a review with an emphasis on human function. *J. Nutr. Biochem.*, **5**, 58–65.

Arora, A., Nair, M.G. and Strasburg, G.M. (1998) Structure–activity relationships for antioxidant activities of a series of flavonoids in a liposomal system. *Free Radical Biol. Med.*, **24**, 1355–1163.

Aruoma, O.I., Halliwell, B., Aesschbach, R. and Loligers, J. (1992) Antioxidant and pro-oxidant properties of active rosemary constituents: carnosol/carnosic acid. *Xenobiotica*, **22**, 257–268.

Avila, M.A., Velasco, J. A., Cansado, J. and Notario, V. (1994) Quercetin mediates the down regulation of mutant p53 in the human breast cancer cell line MDA–MB468. *Cancer Res.*, **54**, 2424–2428.

Aviram, M. and Fuhrman, B. (1998) LDL oxidation by arterial wall macrophages depends on the oxidative status in the lipoprotein and in the cells: Role of prooxidants vs. antioxidants. *Molec. & Cell. Biochem.* **188**, 149–159.

Bachem, C.W.B., Speckmann, G.-J., Van Der Linde, P.C.G., Verheggen, F.T.M., Hunt, M.D., Steffens, J.C. and Zabeau, M. (1994) Antisense expression of polyphenol oxidase-genes inhibits enzymatic browning in potato tubers. *Bio/technology*, **12**, 1101–1105.

Barja, G., Lopez-Torres, M., Perez-Campo, R., Rojas, C., Cadenas, S., Prat, J. and Pamplona, R. (1994) Dietary vitamin C decreases endogenous protein oxidative damage, malondialdehyde, and lipid peroxidation and maintains fatty acid unsaturation in the guinea pig liver. *Free Radical Biol. Med.*, **17**, 105–115.

Barnard, D.E., Sampugna, J., Berlin, E., Bhathena, S.J. and Knapka, J.J. (1990) Dietary *trans* fatty acids modulate erythrocyte membrane fatty acyl composition and insulin binding in monkeys. *J. Nutr. Biochem.*, **1**, 190–195.

Barz, W. and Koster, J. (1981) Turnover and degradation of secondary (natural) products. In: *The Biochemistry of Plants*, E.E. Conn, ed., Vol. 7, Academic Press, New York, pp. 35–116.

Barzanti, V., Pregnolato, P., Maranesi, M., Bosi, I., Baracca, A., Solaini, G. and Turchetto, E. (1995) Effect of dietary oils containing graded amounts of 18:3 n-6 and 18:4 n-3 on cell plasma membranes. *J. Nutr. Biochem.*, **6**, 21–26.

Beatty, E.R., O'Reilly, J.D., England, T.G., McAnlis, G.T., Young, I.S., Halliwell, B., Geissler, C.A., Sanders, T.A.B. and Wiseman, H. (2000) Effect of dietary quercetin on oxidative DNA damage in healthy human subjects. *Br. J. Nutr.* (in press).

Belguendouz, L., Fremont, L. and Gozzelino, M.-T. (1998) Interaction of transresveratrol and plasma lipoproteins. *Biochem. Pharmacol.*, **55**, 811–816.

Berlin, E., Bhathena, S.J., Judd, J.J., Clevidence, B.A. and Peters, R.C. (1994) Human erythrocyte membrane fluidity and insulin binding are independent of dietary *trans* fatty acids. *J. Nutr. Biochem.*, **5**, 591–598.

Berlin, E., Bhathena, S.J., Judd, J.T., Nair, P.P., Peters, R.C., Bhagavan, H.N., Ballard-Barbash, R. and Taylor, P.R. (1992) Effects of omega-3 fatty acid and vitamin E supplementation on erythrocyte membrane fluidity, tocopherols, insulin binding, and lipid composition in adult men. *J. Nutr. Biochem.*, **3**, 392–400.

Beyer, R.E. (1994) The role of ascorbate in antioxidant protection of biomembranes: interaction with vitamin E and coenzyme Q. *J. Bioenerg. and Biomembranes*, **26**, 349–358.

Bhathena, S.J., Berlin, E., Revett, K. and Ommaya, A.E.K. (1986) Modulation of erythrocyte insulin receptors by dietary lipid. *Ann. N.Y. Acad Sci.*, **463**, 165–167.

Biagi, P.L., Bordoni, A., Hrelia, S., Celadon, M. and Turchetto, E. (1993) The effect of dietary polyphosphatidylcholine on microsomal delta-6-desaturase activity, fatty acid composition and microviscosity in rat liver under oxidative stress. *J. Nutr. Biochem.*, **4**, 690–694.

Block, G., Patterson, B. and Subar, A. (1992) Fruit, vegetables and cancer prevention: A review of the epidemiological evidence. *Nutr. Cancer.*, **18**, 1–29.

Bors, W. and Saran, M. (1987) Radical scavenging by flavonoid antioxidants. *Free Radical Res. Communs.*, **2**, 289–294.

Boyd, N.F. and McGuire, V. (1991). The possible role of lipid peroxidation in breast cancer risk. *Free Rad. Biol. Med.*, **10**, 185–190.

Bravo, L. (1998) Polyphenols: Chemistry, dietary sources, metabolism and nutritional significance. *Nutr. Rev.*, **56**, 317–333.

Brown, J.E., Khodr, H., Hider, R.C. and Rice-Evans, C.A. (1998) Structural dependence of flavonoid interactions with Cu^{2+} ions: implications for their antioxidant properties. *Biochem. J.*, **330**, 1173–1178.

Buettner, G.R. (1993) The pecking order of free radicals and antioxidants: lipid peroxidation, α-tocopherol and ascorbate. *Arch. Biochem. Biophys.*, **300**, 535–543.

Burns, D.J. and Bell, R.M. (1992) Lipid regulation of protein kinase C. In: *Protein kinase C: Current Concepts and Future Prospectives*, D.S. Lester and R.M. Epand eds, Ellis Horwood, Chichester, pp. 25–40.

Burns, C.P. and Spector, A.A. (1994) Biochemical effects of lipids on cancer therapy *J. Nutr. Biochem.*, **5**, 114–123.

Burri, J., Graf, M., Lambelet, P. and Loliger, J. (1989) Vanillin: More than a flavouring agent – a potent antioxidant. *J. Sci. Food Agric.*, **48**, 59–56.

Burton, G.W. and Ingold, K.U. (1984) β-Carotene: an unusual type of lipid antioxidant. *Science* **224**, 569–573.

Cadenas, E. and Sies, H. (1998) The lag phase. *Free Radical Res.*, **28**, 601–609.

Cao, G., Booth, S.L., Sadowski, J.A. and Prior, R.L. (1998) Increases in human plasma antioxidant capacity after consumption of controlled diets high in fruit and vegetables. *Am. J. of Clin. Nutr.*, **68**, 1081–1087.

Carson, C.A., Caggiula, A.W., Meilahn, E.N., Matthews, K.A. and Kuller, L.H. (1994) Coffee consumption: relationship to blood lipids in middle-aged women. *Int. J. Epidemiol.*, **23**, 523–527.

Cerutti, P. (1994) Oxy-radicals and cancer. *Lancet* **344**, 862–863.

Cheeseman, K.H. (1993) Lipid peroxidation, and cancer. In: *DNA and Free Radicals*, B. Halliwell and O.I. Aruoma, eds. Ellis Horwood, Chichester, pp. 109–144.

Clarke, R., van den Berg, H.W. and Murphy, R.F. (1990) Reduction of the membrane fluidity of human breast cancer cells by tamoxifen and 17β-estradiol. *J. Natl Cancer Inst.*, **82**, 1702–1705.

Colston, K.W., MacKay, A.G., James, S.Y., Binderup, L., Chander, S. and Coombes, C. (1992) EB1089: a new vitamin D analogue that inhibits the growth of breast cancer cell *in vivo* and *in vitro*. *Biochem. Pharmacol.*, **44**, 2273–2280.

Cook, N.C. and Samman, S. (1996) Flavonoids: Chemistry, metabolism, cardioprotective effects and dietary sources. *J. Nutr. Biochem.*, **7**, 66–76.

Croft, K.D., Williams, P., Dimmitt, S., Abu-Amsha, R. and Beilin, L.J. (1995) Oxidation of low-density lipoproteins: effect of antioxidant content, fatty acid composition and intrinsic phospholipase activity on susceptibility to metal ion-induced oxidation. *Biochim. Biophys. Acta*, **1254**, 250–256.

Cross, C.E., Van der Vliet, A., Louise, S., Thiele, J.J. and Halliwell, B. (1998) Oxidative stress and antioxidants at biosurfaces: Plants, skin, and respiratory tract surfaces. *Environ. Health Perspect.*, **106** (Suppl. 5), 1241–1251.

Darley-Usmar, V., Wiseman, H. and Halliwell, B. (1995) Nitric oxide and oxygen radicals: a question of balance. *FEBS Lett.*, **369**, 131–135.

Dean, R.T., Gieseg, S. and Davies, M.J. (1993) Reactive species and their accumulation on radical-damaged proteins. *Trends Biochem. Sci.*, **18**, 437–441.

Dean, T.S., O'Reilly, J., Bowey, E., Wiseman, H., Rowland, I. and Sanders, T.A.B. (1998) The effects of soyabean isoflavones on plasma HDL concentrations in healthy male and female subjects. *Proc. Nutr. Soc.*, **57**, 123A.

Decker, E.A. (1997) Phenolics: Prooxidants or antioxidants? *Nutr. Rev.*, **55**, 396–398.

De Whalley, C.V., Rankin, S.M., Hoult, J.R.S., Jessup, W. and Leake, D.S. (1990) Flavonoids inhibit the oxidative modification of low density lipoproteins by macrophages. *Biochem. Pharmacol.*, **39**, 1743–1750.

Diplock, A.T., Rice-Evans, C. A. and Burdon, R.H. (1994) Is there a significant role for lipid peroxidation in the causation of malignancy and for antioxidants in cancer prevention. *Cancer Res.* (suppl), **54**, 1952s–1956s.

Dmitriev, L.F., Ivanova, M.V. and Lankin, V.Z. (1994) Interaction of tocopherol with peroxyl radicals does not lead to the formation of lipid hydroperoxides in liposomes. *Chem. and Phys. of Lipids*, **69**, 35–39.

Dumont, J.E., Jauniaux, J.C. and Rogers, P.P. (1989) The cyclic AMP-mediated stimulation of cell proliferation. *Trends Biochem. Sci.*, **14**, 67–71.

Dutta-Roy, A.K., Gordon, M.J., Campbell, F.M., Duthie, G.G. and James, W.P.T. (1994) Vitamin E requirements, transport and metabolism: Role of α-tocopherol-binding proteins. *J. Nutr. Biochem.*, **5**, 562–570.

Dziedzic, S.Z. and Hudson, B.J.F. (1983) Polyhydroxy chalcones and flavanones as antioxidants for edible oils. *Food Chem.*, **12**, 205–212.

Dziedzic, S.Z., Hudson, B.J.F. and Barnes, G. (1985) Polyhydroxydihydrochalcones as antioxidants for lard. *J. Agric. Food Chem.*, **33**, 244–246.

Eberhardt, M.V., Lee, C.Y. and Liu, R.H. (2000) Antioxidant activity of fresh apples. *Nature*, **405**, 903–904.

Frei, B., Kim, M.C. and Ames, B.N. (1990) Ubiquinol-10 is an effective lipid-soluble antioxidant at physiological concentrations. *Proc. Natl. Acad. Sci. USA*, **87**, 4879–4883.

Fruebis, J., Bird, D.A., Pattison, J. and Palinski, W. (1997) Extent of antioxidant protection of plasma LDL is not a predictor of the antiatherogenic effect of antioxidants. *J. Lipid Res.*, **38**, 2455–2464.

Furhrman, B., Lavy, A. and Aviram, M. (1995) Consumption of red wine with meals reduces the susceptibility of human plasma and low-density lipoprotein to lipid peroxidation. *Am. J. Clin. Nutr.*, **61**, 549–554.

Gamou, S. and Shimizu, N. (1995) Hydrogen peroxide preferentially enhances the tyrosine phosphorylation of epidermal growth factor receptor. *FEBS Lett.*, **357**, 161–164.

Garcia-Olmedo, F., Pineiro, M. and Diaz, I. (1994) Dances to a redox tune. *Plant Molec. Biol.*, **26**, 11–13.

Goetghebeur, M. and Kermasha, S. (1996) Inhibition of polyphenol oxidase by copper-metallothionein from *Aspergillus niger*. *Phytochemistry*, **42**, 935–940.

Goldberg, I. (ed) (1994) *Functional Foods, Designer Foods, Pharmafoods, Nutraceuticals.* Chapman and Hall, New York.

Goldring, C.E.P., Rice-Evans, C.A., Burdon, R.H., Rao, R., Haw, I. and Diplock, A.T. (1993) α-Tocopherol uptake and its influence on cell proliferation and lipid peroxidation in transformed and non-transformed baby hamster kidney cells. *Arch. Biochem. Biophys.*, **303**, 429–435.

Goss, S.P.A., Kalyanaraman, B. and Hogg, N. (1998) Antioxidant effects of nitric oxide and nitric oxide donor compounds on low-density lipoprotein oxidation. *Methods in Enzymol.*, **301**, 444–453.

Gschwent, M., Kittstein, W. and Marks, F. (1991) Protein kinase C activation by phorbol esters: do cysteine-rich regions and pseudosubstrate motifs play a role? *Trends Biochem. Sci.*, **16**, 167–169.

Gurr, M.I. and Harwood, J.L. (1991) *Lipid Biochemistry: An Introduction*, 4th edn. Chapman and Hall, London.

Haffner, S.M. and Valdez, R.A. (1995) Endogenous sex hormones: impact on lipids lipoproteins, and insulin. *Am. J. Med.*, **98**, 40S–47S.

Halliwell, B. (1994) Free radicals and antioxidants: a personal view. *Nutr. Rev.*, **52**, 253–265.

Halliwell, B. (1995a) Antioxidant characterization and mechanism. *Biochem. Pharmacol.*, **49**, 1341–1348.

Halliwell, B. (1995b) Oxidation of low-density lipoproteins: questions of initiation, propagation, and the effect of antioxidants. *Am. J. Clin. Nutr.*, **61** (suppl.), 670S–677S.

Halliwell, B. and Chirico, S. (1993) Lipid peroxidation: Its mechanism, measurement and significance. *Am. J. Clin. Nutr.*, **57**, 1S–25S.

Halliwell, B. and Gutteridge, J.M.C. (1999) *Free Radicals in Biology and Medicine*, 3rd edn. Clarendon Press, Oxford.

Heller, F.R., Descamps, O. and Hondekijn, J.-C. (1998) LDL oxidation: Therapeutic perspectives. *Atherosclerosis*, **137** (suppl.), S25–S31.

Hertog, M.G., Feskens, E.J., Hollman, P.C., Katan, M.B. and Kromhout, D. (1993) Dietary antioxidant flavonoids and risk of coronary heart disease. *Lancet*, **342**, 1007–1011.

Hertog, M.G.L. and Hollman, P.C.H. (1996) Potential health effects of the dietary flavonol quercetin. *Eur. J. Clin. Nutr.*, **50**, 63–71.

Hertog, M.G.L., Hollman, P.C., Katan, M.B. and Kromhout, D. (1993) Intakes of potentially anticarcinogenic flavonoids and their determinants in adults in The Netherlands. *Nutr. Cancer*, **20**, 21–29.

Hertog, M.G.L., Sweetnam, P.M., Fehily, A.M., Elwood, P.C. and Kromhout, D. (1997) Antioxidant flavonols and ischemic heart disease in a Welsh population of men: the Caerphilly Study. *Am. J. Clin. Nutr.*, **65**, 1489–1494.

Hightower, K.R., McCready, J.P. and Borchman, D. (1994) Membrane damage in UV-irradiated lenses. *Photochem. and Photobiol.*, **59**, 485–490.

Hoffman, R.M. and Garewal, H.S. (1995) Antioxidants and the prevention of coronary heart disease. *Arch. Int. Med.*, **155**, 241–246.

Hogg, N. and Kalyanaraman, B. (1998) Nitric oxide and low-density lipoprotein oxidation. *Free Radical Res.*, **28**, 593–600.

Holting, T., Siperstein, A.E., Clark, O.H. and Duh, Q.Y. (1995) Epidermal growth factor (EGF) and transforming growth factor α-stimulated invasion and growth of follicular thyroid cancer cells can be blocked by antagonism to the EGF receptor and tyrosine kinase *in vitro*. *Eur. J. Endocrinol.*, **132**, 229–235.

Houslay, M.D. (1985) Regulation of adenylate cyclase (EC4.6.1.1.) activity by its lipid environment. *Proc. Nutr. Soc.*, **44**, 157–165.

Hruszkewycz, A.M. (1992) Lipid peroxidation and mtDNA degeneration. A hypothesis. *Mutation Res.*, **275**, 243–248.

Hubbard, S.R., Wei, L., Ellis, L. and Hendrickson, W.A. (1994) Crystal structure of the tyrosine kinase domain of the human insulin receptor. *Nature*, **372**, 746–753.

Huber, L.A., Scheffler, E., Poll, T., Ziegler, R. and Dresel, H.A. (1990) 17 Beta-estradiol inhibits LDL oxidation and cholesteryl ester formation in cultured macrophages. *Free Radical Res. Communs.*, **8**, 167–173.

Hunter, D.J., Manson, J.E., Colditz, G.A., Stampfer, M.J., Rosner, B., Hennekens, C.H., Speizer, F.E. and Willet, W.C. (1993) A prospective study of the intake of vitamins C, E and A and the risk of breast cancer. *N. Engl. J. Med.*, **329**, 234–240.

Ioka, K., Tsushida, T., Takei, Y., Nakatani, N. and Terao, J. (1995) Antioxidative activity of quercetin monoglycosides in solution and phospholipid bilayers. *Biochim. Biophys. Acta*, **1234**, 99–104.

Imai, K. and Nakachi, K. (1995) Cross sectional study of effects of drinking green tea on cardiovascular and liver diseases. *BMJ*, **310**, 693–696.

Jacobs, T.W. (1995) Cell cycle control. *Annu. Rev. Plant Physiol. Plant. Mol. Biol.*, **46**, 317–339.

James, D.J., Passey, A.J. and Baker, S.A. (1994) Stable gene expression in transgenic apple tree tissues and segregation of transgenes in the progeny – preliminary evidence. *Euphytica*, **77**, 119–121.

Jansson, C., Harmala, A.S., Torvola, D.M. and Slotte, J.P. (1993) Effects of the phospholipid environment in the plasma membrane on receptor interaction with the adenylyl cyclase complex of intact cells. *Biochim. Biophys. Acta*, **1145**, 311–319.

Jenner, P. (1994) Oxidative damage in neurodegenerative disease. *Lancet*, **344**, 796–798.

Joubert, P., Sangwan, R.S., El Arabi Aouad, M., Beaupere, D. and Sangwan-Norrel, B.S. (1995) Influence of phenolic compouns on *Agrobacterium vir* gene induction and onion gene transfer. *Phytochemistry*, **40**, 1623–1628.

Jovanovic, S.V., Steenken, S., Tosic, M., Marjanovic, B. and Simic, M.G. (1994) Flavonoids as antioxidants. *J. Am. Chem. Soc.*, **116**, 4846–4851.

Kagan, V.E., Bakalova, R.A., Koynova, G.M., Tyurin, V.A. Serbinova, E.A., Petkov, V.V., Petkov, V.D., Staneva, D.S. and Packer, L. (1992) Antioxidant protection of the brain against oxidative stress. In: *Free Radicals in the Brain: Aging, Neurological and Mental Disorders*, L. Packer, L. Prilipko and Y. Christen, eds, Springer-Verlag, Berlin, pp. 49–61.

Kagan, V.E., Serbinova, E.A., Bakalova, R.A., Stoytchev, T.S., Erin, A.N., Prilipko, L.L. and Evstigneeva, R.P. (1990a) Mechanisms of stabilization of biomembranes by alpha-tocopherol. *Biochem. Pharmacol.*, **40**, 2403–2413.

Kagan, V., Serbinova, E. and Packer, L. (1990b) Antioxidant effects of ubiquinones in microsomes and mitochondria are mediated by tocopherol recycling. *Biochem. Biophys. Res. Communs.*, **169**, 851–857.

Kaneko, T., Kaji, K. and Matsuo, M. (1994) Protection of linoleic acid and hydroperoxide-induced cytotoxicity by phenolic antioxidants. *Free Radical Biol. and Med.*, **16**, 405–409.

Kayyali, R., Marriott, C. and Wiseman, H. (1994) Tamoxifen decreases drug efflux from liposomes. Relevance to its ability to reverse multidrug resistance in cancer cells. *FEBS Lett.*, **344**, 221–224.

Knight, J.A. (1998) Free radicals: Their history and current status in aging and disease. *Ann Clin. & Lab. Sci.*, **28**, 331–346.

Kolesnick, R. and Golde, D.W. (1994) The spingomyelin pathway in tumor necrosis factor and interleukin-1 signaling. *Cell*, **77**, 325–328.

Kuno, T., Hozumi, M., Morinobu, T., Murata, T., Mingci, Z. and Tamai, H. Antioxidant vitamin levels in plasma and low density lipoprotein of obese girls. *Free Radical Res.*, **28**, 81–86.

Lacort, M., Leal, A.M., Liza, M., Martin, C., Martinez, R. and Ruiz-Larrea, M.B. (1995) Protective effect of estrogens and catechol estrogens against peroxidative membrane damage *in vitro*. *Lipids*, **30**, 141–146.

Laranjinha, J.A., Almeida, L.M. and Madeira, V.M. (1994) Reactivity of dietary phenolic acids with peroxyl radicals: antioxidant activity upon low density lipoprotein peroxidation. *Biochem. Pharmacol.*, **48**, 487–494.

Laughton, M.J., Halliwell, B., Evans, P.J. and Hoult, R.J.S. (1989) Antioxidant and pro-oxidant actions of the plant phenolics quercetin, gossypol and myricetin. *Biochem. Pharmacol.*, **38**, 2859–2865.

Lea, A.G.H. (1984) Tannins and colours in English cider apples. *Fluessiges-obst.*, **51**, 356–361.

Lea, A.G.H. (1995) Enzymes in the production of beverages and fruit juices. In: *Enzymes in Food Processing*, G.A. Tucker and L.E.J. Woods, eds, Blackie, London.

Leonessa, F., Jacobson, M., Boyle, B., Lippmann, J., McGarvey, M. and Clarke, R. (1994) Effect of tamoxifen on the multidrug-resistant phenotype in human breast cancer cells: isobologram, drug accumulation, and M_r 170,000 glycoprotein (gp170) binding studies. *Cancer Res.*, **54**, 441–447.

Lewin, G. and Popov, I. (1994) Antioxidant effects of aqueous garlic extract. 2nd communication: Inhibition of the $Cu(2^+)$-initiated oxidation of low density lipoproteins. *Arzneimittel-Forschung*, **44**, 604–607.

Lipkin, M. and Boone, C. W. (1991) Calcium, vitamin D and cancer. *Cancer Res.*, **51**, 3069–3070.

Liu, M., Pelling, J.C., Ju, J., Chu, E. and Brash, D.E. (1998) Antioxidant action via p53-mediated apoptosis. *Cancer Res.*, **58**, 1723–1729.

Lloyd, T., Chinchilli, V.M., Rollings, N., Kieselhorst, K., Tregea, D.F., Henderson, N.A. and Sinoway, L.I. (1998) Fruit consumption, fitness, and cardiovascular health in female adolescents: The Penn State Young Women's Health Study. *Amer. J. Clin. Nutr.*, **67**, 624–630.

Loft, S. and Poulsen, H.E. (1998) Markers of oxidative damage to DNA: Antioxidants and molecular damage. *Methods in Enzymol.*, **300**, 166–184.

Mangiapane, H., Thomson, J., Salter, A., Brown, S., Bell, G.P. and White, D.A. (1992) The inhibition of the oxidation of low density lipoproteins by (+)-catechin, a naturally occurring flavonoid. *Biochem. Pharmacol.* **43**, 445–450.

Mann, J. (1992) *Secondary Metabolism*. Clarendon Press, Oxford.

Mathiesen, L., Wang, S., Halvorsen, B., Malterud, K.E. and Surd, R.B. (1996) Inhibition of lipid peroxidation in low-density lipoprotein by the flavonoid myrigalone B and ascorbic acid. *Biochem. Pharmacol.*, **51**, 1719–1726.

McCallum, C.D. and Epand, R.M. (1995) Insulin receptor autophosphorylation and signalling is altered by modulation of membrane physical properties. *Biochemistry*, **34**, 1815–1824.

Mecocci, P., MacGarvey, U. and Beal, M.F. (1995) Oxidative damage to mitochondrial DNA is increased in Alzheimer's disease. *Annals Neurol.*, **36**, 747–751.

Messina, M.J., Persky, V., Setchell, K.D. and Barnes, S. (1994) Soy intake and cancer risk: A review of the *in vitro* and *in vivo* data. *Nutr. and Cancer*, **21**, 113–131.

Miksicek, R.J. (1993) Commonly occurring plant flavonoids have estrogenic activity. *Molec. Pharmacol.*, **44**, 37–43.

Miksicek, R.J. (1994) Interaction of naturally occurring nonsteroidal estrogens with expressed recombinant human estrogen receptor *J. Steroid Biochem. Molec. Biol.*, **49**, 153–160.

Miksicek, R.J. (1995) Estrogenic flavonoids: structural requirements for biological activity. *Proc. Soc. Experimental Biol. Med.*, **208**, 44–50.

Miller, N.J., Diplock, A.T. and Rice-Evans, C.A. (1995) Evaluation of the total antioxidant activity as a marker of the deterioration of apple juice on storage. *J. Agric. Food Chem.*, **43**, 1794–1801.

Miller, N.J. and Rice-Evans, C.A. (1997) The relative contributions of ascorbic acid and phenolic antioxidants to the total antioxidant activity of orange and apple fruit juices and a blackcurrant drink. *Food Chem.* **60**, 331–337.

Morrissey, P.A., Quinn, P.B. and Sheehy, P.J.A. (1994) Newer aspects of micronutrients in chronic disease: Vitamin E. *Proc. Nutr. Soc.*, **53**, 571–582.

New, R.R.C. (1992) *Liposomes: A Practical Approach*. IRL Press, Oxford.

Nicolas, J.J., Richard-Forget, F.C., Goupy, P.M., Amiot, M.-J. and Aubert, S.Y. (1994) Enzymatic browning reactions in apple and apple products. *Crit. Rev. Food Sci. Nutr.*, **34**, 109–157.

Nielsen, L. B. (1996) Transfer of low density lipoprotein into the arterial wall and role of atherosclerosis. *Atherosclerosis*, **123**, 1–16.

Noguchi, N. and Niki, E. (1998) Dynamics of vitamin E action against LDL oxidation. *Free Radical Res.*, **28**, 561–572.

Nordmann, R. (1994) Alcohol and antioxidant systems. *Alcohol and Alcoholism*, **29**, 513–522.

Nyyssonen, K., Porkkala, E., Salonen, R., Korpela, H. and Salonen, J.T. (1994) Increase in oxidation resistance of atherogenic serum lipoproteins following antioxidant supplementation; a randomized double-blind placebo-controlled clinical trial. *Eur. J. Clin. Nutr.*, **48**, 633–642.

Ohyashiki, T., Kumada, Y., Hatanaka, N. and Matsui, K. (1994a) Oxygen radical-induced inhibition of alkaline phosphatase activity in reconstituted membranes. *Arch. Biochem. Biophys.*, **313**, 310–317.

Ohyashiki, T., Sakata, N. and Matsui, K. (1994b) Changes in SH reactivity of the protein in porcine intestinal brush-border membranes associated with lipid peroxidation. *J. of Biochem.*, **115**, 224–229.

Oliver, C.N., Ahn, B.-W., Moerman, E.J., Goldstein, S. and Stadtman, E.R. (1987) Age-related changes in oxidized proteins. *J. Biol. Chem.*, **262**, 5488–5491.

Olivieri, O., Stanzial, A.M., Girelli, D., Trevisan, M.T., Guarini, P., Terzi, M., Caffi, S., Fontana, F., Casaril, M., Ferrari, S. and Corrocher, S. (1994) Selenium status, fatty acids, vitamins A and E and aging: the Nove study. *Am. J. Clin. Nutr.*, **60**, 510–517.

O'Reilly, J.D., Sanders, T.A.B. and Wiseman, H. (2000) Flavonoids protect against oxidative damage to LDL *in vitro*: use in selection of a flavonoid rich diet and relevance to LDL oxidation resistance *ex vivo*? *Free Radical Research* (in press).

Ortega, R.M., Lopez-Sobaler, A.M., Quintas, M.E., Martinez, R.M. and Andres, P. (1998) Influence of smoking on vitamin C status during the third trimester of pregnancy and on vitamin C levels in maternal milk. *J. Amer. Coll. Nutr.*, **17**, 379–384.

Parfitt, V.J., Rubba, P., Bolton, C., Marotta, G., Hartog, M. and Mancini, M. (1994) A comparison of antioxidant status and free radical peroxidation of plasma lipoproteins in healthy young persons from Naples and Bristol. *Eur. Heart J.*, **15**, 871–876.

Parshad, R., Sanford, K.K., Price, F.M., Steele, V.E., Tarone, R.E., Kelloff, G.J. and Boone, C.W. (1998) Protective action of plant polyphenols on radiation-induced chromatid breaks in cultured human cells. *Anticancer Res.*, **18**, 3263–3266.

Parthasarathy, S., Auge, N. and Santanam, N. (1998) Implications of lag time concept in the oxidation of LDL. *Free Radical Res.*, **28**, 583–891.

Pasman, P.C. and Schouten, H.C. (1993) Multidrug resistance mediated by P-glycoprotein in haematological malignancies. *Netherlands J. Med.*, **42**, 218–231.

Peck, M.D. (1994) Interactions of lipids with immune function I: Biochemical effects of dietary lipids on plasma membranes. *J. Nutr. Biochem.*, **5**, 466–478.

Pellett, L.J., Anderson, H.J., Chen, H. and Tappel, A.L. (1994) β-Carotene alters vitamin E protection against heme protein oxidation and lipid peroxidation in chicken liver slices. *J. Nutr. Biochem.*, **5**, 479–484.

Peterson, J. and Dwyer, J. (1998) Flavonoids: Dietary occurrence and biochemical activity. *Nutr. Res.*, **18**, 1995–2018.

Pietta, P. and Simonetti, P. (1998) Dietary flavonoids and interaction with endogenous antioxidants. *Biochem. & Molec. Biol. Int.*, **44**, 1069–1074.

Poulsen, H.E., Prieme, H. and Loft, S. (1998) Role of oxidative DNA damage in cancer initiation and promotion. *Eur. J. Cancer Prev.*, **7**, 9–16.

Powis, G. (1989) Free radical formation by antitumor quinones. *Free Radical Biol. Med.*, **6**, 63–101.

Rakwal, R., Hasegwa, O. and Kodoma, O. (1996) A methyltransferase for synthesis of the flavanone phytoalexin sakuranetin in rice leaves. *Biochem. Biophys. Res. Commun.*, **222**, 732–735.

Reddy, A.C.P. and Lokesh, B.R. (1994) Alterations in lipid peroxides in rat liver by dietary n-3 fatty acids: modulation of antioxidant enzymes by curcumin, eugenol, and vitamin E. *J. Nutr. Biochem.*, **5**, 181–188.

Renaud, S. and De Lorgeril, M. (1992) Wine, alcohol platelets and the French paradox for coronary heart disease. *Lancet*, **339**, 1523–1526.

Rice-Evans, C.A., Miller, N.J., Bolwell, P.G., Bramley, P.M. and Pridham, J.B. (1995) The relative antioxidant activities of plant-derived polyphenolic flavonoids. *Free Radical Res.*, **22**, 375–383.

Rice-Evans, C.A., Miller, N.J. and Paganga, G. (1996) Structure–antioxidant activity relationships of flavonoids and phenolic acids. *Free Radical Biol. Med.*, **20**, 933–956.

Ridgway, T., O'Reilly, J., West, G., Tucker, G. and Wiseman, H. (1996a) Potent antioxidant properties of novel apple-derived flavonoids and commercial potential as food additives. *Biochem. Soc. Trans.*, **24**, 391S.

Ridgway, T., O'Reilly, J., West, G., Tucker, G. and Wiseman, H. (1997) Antioxidant action of novel derivatives of the apple-derived flavonoid phloridzin compared to oestrogen: relevance to potential cardioprotective action. *Biochem. Soc. Trans.*, **25**, 106S.

Ridgway, T.J. and Tucker, G.A. (1997a) Apple: a new agrochemical crop. *Biochem. Soc. Trans.*, **25**, 110S.

Ridgway, T.J. and Tucker, G. A. (1997b) Phloridzin derivatives: food additives/ chemopreventative drugs of the future. *Biochem. Soc. Trans.*, **25**, 109S.

Ridgway, T.J. and Tucker, G.A. (1997c) Prospects for the production and use of new improved dietary oestrogens for cardioprotection. *Biochem. Soc. Trans.* Invited Colloquium paper, **25**, 59–63.

Rifici, V.A. and Khachadurian, A.K. (1992) The inhibition of low-density lipoprotein oxidation by 17-β estradiol. *Metabolism*, **41**, 1110–1114.

Rikans, L.E. and Hornbrook, K. R. (1997) Lipid peroxidation, antioxidant protection and aging. *Biochim. Biophy. Acta – Molecular Basis of Disease*, **1362**, 116–127.

Ruiz-Larrea, M.B., Leal, A.M., Liza, M., Lacort, M. and de Groot, H. (1994) Antioxidant effects of estradiol and 2-hydroxyestradiol on iron-induced lipid peroxidation in rat liver microsomes. *Steroids*, **59**, 383–388.

Rosin, M.P., Anwar, W.A. and Ward, A.J. (1994) Inflammation, chromosomal instability and cancer: the Schistosomiasis model. *Cancer Res.* (Suppl.) **54**, 1929s–1933s.

Sack, M.N., Rader, D.J. and Cannon, R.O. (1994) Oestrogen and inhibition of oxidation of low-density lipoproteins in postmenopausal women. *Lancet*, **343**, 269–270.

Saito, M. and Nakatsugawa, K. (1994) Increased susceptibility of liver to lipid peroxidation after ingestion of a high fish oil diet. *Int. J. for Vitamin and Mineral Res.* **64**, 144–151.

Salah, N., Miller, N.J., Paganga, G., Tijburg, L., Bolwell, G.P. and Rice-Evans, C. (1995) Polyphenolic flavanols as scavengers of aqueous phase radicals and as chain-breaking antioxidants. *Arch. Biochem. Biophys.*, **322**, 339–346.

Sandhir, R., Julka, D. and Gill, K.D. (1994) Lipoperoxidative damage on lead exposure in rat brain and its implications on membrane bound enzymes. *Pharmacol. & Toxicol.*, **74**, 66–71.

Schwenke, D.C. (1998) Antioxidants and atherogenesis. *J. Nutr. Biochem.*, **9**, 424–445.

Sestili, P., Guidarelli, A., Dacha, M. and Cantoni, O. (1998) Quercetin prevents DNA single strand breakage and cytotoxicity caused by *tert*-butylhydroperoxide: Free radical scavenging versus iron chelating mechanism. *Free Radical Biol. & Med.*, **25**, 196–200.

Seydel, J.K., Coats, E.A., Cordes, H.P. and Wiese, M. (1994) Drug membrane interaction and the importance for drug transport, distribution, accumulation, efficacy and resistance. *Archiv der Pharm.*, **327**, 601–610.

Shigenaga, M.K., Hagen, T.M. and Ames, B.N. (1994) Oxidative damage and mitochondrial decay in aging. *Proc. Natl. Acad. Sci. USA*, **91**, 10771–10778.

Shimoi, K., Masuda, S., Furugori, M. and Kinae, N. (1994) Radioprotective effect of antioxidative flavonoids in gamma-ray irradiated mice. *Carinogenesis*, **15**, 2669–2672.

Simon, J.A., Hudes, E.S. and Browner, W.S. (1998) Serum ascorbic acid and cardiovascular disease prevalence in U.S. adults. *Epidemiology*, **9**, 316–321.

Smith, C., Mitchinson, M., Aruoma, O.I. and Halliwell, B. (1992) Stimulation of lipid peroxidation and hydroxyl radical generation by the contents of human atherosclerotic lesions. *Biochem. J.*, **286**, 901–905.

Sohal, R.S., Ku, H.-H., Agarwal, S., Forster, M.J. and Lal, H. (1994) Oxidative damage, mitochondrial oxidant generation and antioxidant defenses during aging in response to food restriction in the mouse. *Mech. of Ageing and Dev.*, **74**, 121–133.

Spiteller, G. (1998) Linoleic acid peroxidation – the dominant lipid peroxidation process in low density lipoprotein – and its relationship to chronic diseases. *Chem. and Phys. Lipids*, **95**, 105–162.

Stadtman, E.R. (1993) Oxidation of free amino acids and amino acid residues in proteins by radiolysis and by metal-catalysed reactions. *Ann. Rev. Biochem.*, **62**, 797–821.

Stadler, R.H., Turesky, R.J., Muller, O., Markovic, J. and Leong-Morgenthaler, P.M. (1994) The inhibitory effects of coffee on radical-mediated oxidation and mutagenicity. *Mutat. Res.*, **308**, 177–190.

Stah, W., Junghans, A., De Boer, B., Driomina, E.S., Briviba, K. and Sies, H. (1998) Carotenoid mixtures protect multilamellar liposomes against oxidative damage synergistic effects of lycopere and lutein. *FEBS Lett.*, **427**, 305–308.

Sugioka, K., Shimosegawa, M. and Nakano, M. (1987) Estrogens as natural antioxidants of membrane phospholipid peroxidation. *FEBS Lett.*, **210**, 192–194.

Takeda, Y., Nishio, K. and Saijo, N. (1994) Reversal of multidrug resistance by tyrosine-kinase inhibitors in a non-P-glycoprotein-mediated multidrug-resistant cell line. *Int. J. Cancer*, **57**, 229–239.

Tappel, A. (1998) Models of antioxidant protection against biological oxidative damage. *Lipids*, **33**, 947.

Terao, J., Piskula, M. and Yao, Q. (1994) Protective effect of epicatechin, epicatechin gallate and quercetin on lipid peroxidation in phospholipid bilayers. *Arch. Biochem. Biophys.*, **308**, 278–284.

Tertov, V.V., Sobenin, I.A., Kaplun, V.V. and Orekhov, A.N. (1998) Antioxidant content in low density lipoprotein and lipoprotein oxidation *in vivo* and *in vitro*. *Free Radical Res.*, **29**, 165–173.

Thomas, J.P., Kalyanaraman, B. and Girotti, A.W. (1994) Involvement of pre-existing lipid hydroperoxides in Cu^{2+}-stimulated oxidation of low-density lipoprotein. *Arch. Biochem. Biophys.*, **315**, 244–254.

Toyokuni, S., Okamoto, K., Yodoi, J. and Hiai, H. (1995) Persistent oxidative stress in cancer. *FEBS Letters*, **358**, 1–3.

Van Ginkel, G. and Sevanian, A. (1994) Lipid peroxidation-induced membrane structural alterations. *Methods in Enzymology*, **233**, 273–288.

Vedavanam, K., Srijayanta, S., O'Reilly. J., Raman, A. and Wiseman, H. (1999) Antioxidant action and potential antidiabetic properties of an isaflovonoid-containing soyabean phytochemical extract (SPE). *Phytother. Res.*, **13**, 601–608.

Vinson, J.A. and Hontz, B.A. (1995) Phenol antioxidant index: comparative effectiveness of red and white wines. *J. Agric. Food. Sci.* **43**, 401–403.

Viret, J., Daveloose, D. and Leterrier, F. (1990) Modulation of the activity of functional membrane proteins by the lipid bilayer fluidity. In *Membrane Transport and Information Storage*. R.C. Aloia, C.C. Curtain and L.M. Gordon, eds. Wiley-Liss, New York, pp. 239–253.

Vita, J.A., Keaney, J.F., Jr, Raby, K.E., Morrow, J.D., Freedman, J.E., Lynch, S. Koulouris, S.N., Hankin, B.R. and Frei, B. (1998) Low plasma ascorbic acid independently predicts the presence of an unstable coronary syndrome. *J. Amer. Coll. Cardiol.*, **31**, 980–986.

Wagner, B.A., Beuttner, G.R. and Burns, C.P. (1993) Increased generation of lipid-derived and ascorbate free radicals by L1210 cells exposed to the ether lipid edelfosine. *Cancer Res.*, **53**, 711–713.

Wagner, B.A., Beuttner, G.R. and Burns, C.P. (1992) Membrane peroxidative damage by the ether lipid class of antineoplastic agents. *Cancer Res.*, **52**, 6045–6051.

Walker, J.R.L. and Ferrar, P.H. (1995) The control of enzymic browning in foods. *Chem. and Ind.*, **16 October**, 836–839.

Wang, H., Cao, G. and Prior, R.L. (1996) Total antioxidant capacity of fruits. *J. Agric. Food Chem.*, **44**, 701–705.

Whitehead, T.P., Robinson, D., Allaway, S., Syms, J. and Hale, A. (1995) Effect of red wine ingestion on the antioxidant capacity of serum. *Clin. Chem.*, **41**, 32–35.

Williamson, P. and Schlegel, R.A. (1994) Back and forth: the regulation and function of transbilayer phospholipid movement in eukaryotic cells. *Molec. Membrane Biol.*, **11**, 199–216.

Wiseman, H. (1993) Vitamin D is a membrane antioxidant. Ability to inhibit iron-dependent lipid peroxidation in liposomes compared to cholesterol ergosterol and tamoxifen and relevance to anticancer action. *FEBS Lett.*, **326**, 285–288.

Wiseman, H. (1994a) *Tamoxifen: Molecular Basis of Use in Cancer Treatment and Prevention*. John Wiley, Chichester.

Wiseman, H. (1994b) The antioxidant action of a pure antioestrogen. Ability to inhibit lipid peroxidation compared to tamoxifen and 17 β-oestradiol and relevance to its anticancer potential. *Biochem. Pharmacol.* **47**, 493–498.

Wiseman, H. (1996a) Dietary influences on membrane function: importance in protection against oxidative damage and disease. *J. Nutr. Biochem.*, **7**, 2–15.

Wiseman, H. (1996b) Role of dietary phytoestrogens in the protection against cancer and heart disease. *Biochem. Soc. Trans.* Invited Colloquium paper, **24**, 795–800.

Wiseman, H. and Halliwell, B. (1996) Damage to DNA by reactive oxygen and nitrogen species: role in inflammatory disease and progression to cancer. *Biochem. J.*, **313**, 17–29.

Wiseman, H., Laughton, M.J., Arnstein, H.R.V., Cannon, M. and Halliwell, B. (1990) The antioxidant action of tamoxifen and its metabolites. Inhibition of lipid peroxidation. *FEBS Lett.*, **263**, 192–194.

Wiseman, H., O'Reilly, J.D., Adlercreutz, H., Mallet, A.I., Bowey, G.A., Rowland, I.R. and Sanders, T.A.B. (2000) Isoflavone phytoestrogens consumed in soy decrease F_2-isoprostane concentrations and increase resistance of low-density lipoprotein to oxidation in humans. *Am. J. Clin. Nutr.* (in press).

Wiseman, H., Plitzanopoulou, P. and O'Reilly, J. (1996) Antioxidant properties of ethanolic and aqueous extracts of green tea compared to black tea. *Biochem. Soc. Trans.*, **24**, 390S.

Wiseman, H., Paganga, G., Rice-Evans, C., Halliwell, B. (1993b) Protective actions of tamoxifen and 4-hydroxytamoxifen against oxidative damage to human low-density lipoproteins. A mechanism accounting for the cardioprotective action of tamoxifen? *Biochem. J.*, **292**, 635–638.

Wiseman, H., Quinn, P. and Halliwell, B. (1993) Tamoxifen and related compounds decrease membrane fluidity in liposomes. Mechanism for the antioxidant action of tamoxifen and relevance to its anticancer and cardioprotective actions? *FEBS Lett.*, *330*, 53–56.

Witzum, J.L. (1994) The oxidation hypothesis of atherosclerosis. *Lancet*, **344**, 792–795.

Wolff, S.P. and Dean, R. T. (1987) Glucose autoxidation and protein modification: the potential role of 'autoxidative glycosylation' in diabetes. *Biochem J.*, **245**, 243–250.

Zachowski, A. (1993) Phospholipids in animal eukaryotic membranes: transverse asymmetry and movement. *Biochem. J.*, **294**, 1–14.

Ziouzenkova, O., Sevanian, A., Abuja, P.M., Ramos, P. and Esterbauet, H. (1998) Copper can promote oxidation of LDL by markedly different mechanisms. *Free Radical Biol. Med.*, **24**, 607–623.

2 DNA Injury: Prevention by Antioxidants

BIBRA International, Woodmansterne Road, Carshalton, Surrey SM5 4 DS
Currently: Department of Medical Sciences, University of Bath, Bath BA2 7AY

Most cancer scientists are familiar with the term carcinogen, which describes any chemical or biological molecule that interacts with DNA and can cause toxic, lethal or heritable effects (Brookes 1977). These effects are commonly referred to as genotoxic events since they are ultimately toxic to the genetic material. Carcinogens may interact directly with DNA or may produce other chemicals or molecules, such as free radicals that attack the electron-rich structure of DNA causing breaks in the DNA molecule or the formation of DNA adducts (Miller and Miller 1981). These alterations can have far-reaching consequences such as the inhibition of DNA synthesis or repair, or mutations in chromosomes (Bartsch 1996; Singer 1996).

Oxidative DNA damage

Humans are aerobic organisms and the mitochondria, which produce approximately 80% of our energy requirements, use O_2 as the terminal electron acceptor. This system is a highly efficient but not a 100% effective means of energy production, and a consequence of this is the production of oxygen-derived reactive intermediates that are highly toxic (Troll and Wiesner 1985). Indeed, oxygen itself is a toxic gas, and it is likely that the internal cell environment is subjected to a continual barrage of reactive oxidants as a host of oxygen-derived free radicals are formed (Kehrer 1993), and these can damage cell structures including cellular proteins, carbohydrates, lipids and DNA (Halliwell and Gutteridge 1989). In living cells these reactive metabolites are formed continually as a product of metabolism or other biochemical reactions, or as mediators for important physiological functions (Loft and Poulsen 1996).

Free radicals contain one or more unpaired electrons, and the oxygen free radicals, together with some biologically active non-radical oxygen derivatives are commonly referred to as reactive oxygen species (ROS).

ROS include the superoxide ($O_2^{•-}$), hydroxyl ($OH^•$), peroxyl ($RO_2^•$) and alkoxyl ($RO^•$) free radicals, and the non-radical derivatives (H_2O_2), hypochlorous acid (HOCl) and ozone (O_3) (Halliwell 1994a).

Some cells produce oxidant free radicals deliberately and these reactive products have important physiological functions. Oxidative free radicals play an important role as a part of the body's natural immune system. Phagocytic cells, for example, use superoxide to kill bacteria, fungi (Curnutte and Babior 1987), and virus-infected cells, and to communicate with other neutrophils. Once activated, neutrophils produce large amounts of superoxide from an NADPH oxidase located in the plasma membrane. The superoxide is readily released where it can react with a chemotaxic precursor and attract more neutrophils, thus maintaining the inflammatory momentum (McCord et al. 1982). Other cells, including lymphocytes and fibroblasts, may also use superoxide for intercellular signalling (Burdon and Rice-Evans 1989; Meier et al. 1990) and the activation of signal transduction pathways which may influence the expression of certain growth- and differentiation-related genes (Cerutti et al. 1994). The reactivity of ROS varies depending on the parent molecule and its proximity to other molecules and other free radicals. If a newly generated free radical comes into contact with another free radical, they can react to form a covalent bond, but this does not necessarily mean that the newly formed compound is non-hazardous. When the nitric oxide free radical ($NO^•$) combines with superoxide ($O_2^{•-}$), the reaction product is peroxynitrite ($ONOO^-$) (Huie and Padmaja 1993), which can damage cellular proteins by oxidising sulphhydryl (–SH) groups (Beckman et al. 1994), and breaks down to produce other toxic and reactive products including nitrogen dioxide ($NO_2^•$) and the highly reactive hydroxyl radical ($OH^•$). More often, however, free radicals will react with a non-radical, resulting in a chain reaction which can lead to the formation of a plethora of other radicals including fatty acids which initiate lipid peroxidation (Halliwell 1994a) and the peroxidation of low-density lipoproteins thought to be responsible for the formation of atherosclerotic plaques in cardiovascular disease (Halliwell 1995).

The main chain-propagating reaction in the pro-oxidation of lipids results from the formation of peroxyl radicals (Halliwell and Gutteridge 1989), and the peroxidation of polyunsaturated lipids generates a range of reactive products. These transient species include a number of free radicals and oxygen-centred radicals (alkoxyl and other peroxyl radicals), which are themselves unstable, and can react with lipids to form lipid hydroperoxides that are known to cause damage to DNA (Burcham 1998). Lipid hydroperoxides can form a range of genotoxic substances, and can induce both single- and double-strand breaks in the DNA molecule; lesions which appear to occur most frequently near guanine residues (Inouye 1984).

Exogenous production of oxygen radicals and other reactive species

The hydroxyl radical can be generated in a number of ways. Ionising radiation, to which we are continually exposed, causes fission of the O–H bonds in water generating both hydrogen ($H^\bullet$) and hydroxyl ($OH^\bullet$) free radicals (von Sonntag 1987). Many of the chemical and physical alterations in biomolecules appear to be mediated by free radicals and other reactive intermediates formed following absorption of radiant energy (Greenstock 1993). It is thought that many if not most of the harmful effects of ionizing radiation are due to ionisation of water molecules surrounding DNA which causes the production of hydroxyl radicals close to the DNA molecule (McLennan *et al.* 1980).

An important biochemical pathway thought to be responsible for the production of $OH^\bullet$, is the Haber–Weiss reaction (McLennan *et al.* 1980; Halliwell 1978a,b; McCord and Day 1978), in which superoxide or other reducing agents cause the reduction of Fe^{3+} to Fe^{2+} which is then available to cause the breakdown of H_2O_2 and liberate the hydroxyl radical (Figure 2.1):

$$O_2^{\bullet-} + Fe^{3+} \longrightarrow Fe^+ + O_2$$

$$Fe^{2+} + H_2O_2^- \longrightarrow OH^\bullet + OH^- + Fe^{3+}$$

Figure 2.1. Haber–Weiss reaction. In the presence of metal ions, the highly reactive hydroxyl radical is liberated by the action of superoxide

The metal ions required for this reaction to occur are iron, as shown in the above reaction, or copper which appears to be equally effective as iron in producing hydroxyl radicals (Tkeshelashvili *et al.* 1991). However, in healthy individuals most metal ions, such as iron and copper, are bound to transport and storage proteins and therefore only a small amount is free to participate in the reaction. This sequestration of metal ions has been suggested as an inherent antioxidant defence mechanism (Halliwell and Gutteridge 1989), inhibiting not only the Haber–Weiss reaction but also the genotoxicity of lipid hydroperoxides whose reactivity appears to be mainly due to decomposition products formed in the presence of metal ions (Ueda *et al.* 1985; Yang and Schaich 1996).

With regard to oxidative damage to DNA, the superoxide ($O_2^{\bullet-}$) radical does not appear to be directly able to cause DNA strand breaks (Brawn and Fridovich 1981; Rowley and Halliwell 1983) or interact with DNA bases (Aruoma *et al.* 1989). However, it does appear to play a role in DNA damage, probably due to its participation in the formation of the highly reactive $OH^\bullet$ (Mello Filho *et al.* 1984; Fridovich 1986). This is certainly a feasible mechanism, since all the elements required for the production of $OH^\bullet$ by the Haber–Weiss reaction are present in close proximity to DNA; H_2O_2 can easily traverse cell membranes to the nucleus (Halliwell and Gutteridge 1989), and copper ions are thought to be present in chromosomes (Prutz *et al.* 1990), providing the necessary catalyst for the production of hydroxyl radicals. Metal ions in close proximity to DNA are also thought to be present after release in the cell during oxidative stress (Halliwell 1987).

While $O_2^{\bullet -}$ does not appear to interact directly with DNA, there is evidence that $O_2^{\bullet -}$ finds other direct targets in the cell (Fridovich 1986), and these interactions may lead to the production of other DNA damaging species.

The hydroxyl radical is one of the most reactive and damaging of all the free radicals and will cause damage to whatever it is generated next to (von Sonntag 1987). DNA damage by $OH^{\bullet}$ appears to occur in all cells, and may be a significant contributor to age-dependent development of cancer (Halliwell and Gutteridge 1984). The $OH^{\bullet}$ radical can attack any component of the DNA molecule and can give rise to multiple changes; for example strand breaks (Mello Filho and Meueghini 1983), deoxyribose-derived radicals that cause the release of DNA bases resulting in apurinic and apyramidinic sites, chemical alterations to DNA bases, and sugar adducts that can give rise to DNA single-strand breaks when incubated with alkali (alkali-labile sites) (Halliwell and Aruoma 1991).

Endogenous cellular antioxidant defences

To protect cell structures from damage by oxygen free radicals, aerobic cells have developed elaborate antioxidant defences (Fridovich 1978), and these defences appear to be essential to the sustained health of human cells (Halliwell 1994a). A range of enzymes provide a system of antioxidant defences that work in concert, manipulating free radicals to make them inactive (Cerutti *et al.* 1994). Superoxide dismutase (SOD), for instance, are a family of inducible metalloenzymes (one containing copper and one containing zinc), which remove excess $O_2^{\bullet -}$ by assisting in its conversion to O_2 and H_2O_2 (Fridovich 1978; 1983). The hydrogen peroxide is then removed by a number of enzymes including catalase and selenoprotein glutathione peroxidase (Chance *et al.* 1979) which convert H_2O_2 to water and molecular oxygen (Fridovich 1983).

Glutathione (GSH) is a tripeptide thiol which appears to have evolved in early aerobic cells to protect them against oxidative damage (Fehey and Sundquist, 1991), and is known to rapidly react with radicals resulting from $OH^{\bullet}$ attack on DNA (von Sonntag 1987; Fahey 1988).

One of the most important intracellular sources of ROS are in mitochondria, and they are protected against ROS by both enzymatic and non-enzymatic antioxidants. However, despite this dual approach to handling ROS, the mitochondrial DNA (mtDNA) is still subject to severe oxidative damage; more so than nuclear DNA (Richter 1995). Several hypotheses suggest that defective mitochondria, which result from a culmination of increased production of ROS and a decrease in particular antioxidant defences, leads to an increase in mtDNA modification and thus may contribute to, or be responsible for ageing (Richter 1995).

Oxidative DNA damage and cancer

Upon reviewing the epidemiological evidence, it would appear that occupational exposure accounts for fewer human cancers than previously thought (Doll and Peto 1981; Doll 1992; Ames et al. 1995). More weight has been given to the contributions that age and diet have on cancer incidence via the accumulation of genotoxic products that can cause an increase in the levels of genetic damage (Ames et al. 1995).

Experimental studies, both in vitro and in vivo, suggest that oxidative damage to DNA is an important factor in carcinogenesis (Loft and Poulsen 1996) with active oxygen species acting as both tumour promoters and complete carcinogens (Zimmerman and Cerruti 1984; Cerutti 1985), with the potential to mutate cancer-related genes (Cerruti et al. 1994).

Some tumour promoters, such as the phorbol ester phorbol-myristate-acetate (PMA), have been shown experimentally to enhance tumour formation by stimulating superoxide production via an NADPH-dependant oxidase (Troll and Wiesner 1985). Other tumour promoters such as mezerein (which is related to the phorbol esters) (Troll et al. 1982) and indole alkaloids (which are distinct from phorbol esters) (Fujiki et al. 1981), also stimulate the production of superoxide.

Despite extensive repair, oxidatively modified DNA is abundant in human tissue, and is frequently found in tumour tissue (1-200 oxidised nucleosides per 10^5 intact nucleosides) (Loft and Poulsen 1996). It would appear that the number of damaged nucleosides also increases with age, and experimental evidence in rats has shown that the rate of unrepaired oxidative damage increases with age by about 80 residues per day (Fraga et al. 1990). This increase may be a consequence of a reduced capacity for DNA repair that is concomitant with the ageing process (Wei et al. 1993). Products of DNA repair to these nucleosides are excreted in the urine, and correspond to 10 000 repairs to each cell per day (Loft and Poulsen 1996; Lindsay 1996). The most abundant base lesion due to oxidative attack on DNA is the formation of 8-hydroxyguanine (8-OHG), and it is this lesion that is most often measured as an index of oxidative DNA damage (Floyd et al. 1996b). This lesion (8-OHG) is known to result from the attack of OH$^•$

on guanine, and has been found in cells exposed to oxidative stress (Kasai *et al.* 1986; Floyd *et al.* 1986a). 8-OHG is one of the most mutagenic DNA lesions, and leads to lack of base-pairing, misreading of the modified base and adjacent residues (Poulsen *et al.* 1998), frequently resulting in G–T transversions which are commonly found in tumour-relevant genes (Loft and Poulsen 1996).

It is thought that an increase in the production of ROS, or a decrease in antioxidant defences (i.e. an increased level of oxidative stress), may exacerbate many age-related diseases. Increased oxidative stress can arise from either an increase in the production of ROS (resulting endogenously or from an environmental cause), or the depletion of the antioxidant defences, possibly as a consequence of ageing or poor diet. In addition, chronic inflammation, which is accompanied by a burst of free radicals from activated white blood cells, contributes to DNA damage and mutation (Shacter *et al.* 1988), and is a known risk factor for cancer (Weitzman and Gordon 1990). Despite ingenious cellular antioxidant defences, damage to DNA, lipids, proteins and other molecules by ROS still occurs, and it is this excess damage that is thought to contribute to the development of cancer (Dreher and Junod 1996) and other age-related diseases including cardiovascular disease (Witztum 1994) and neurodegenerative diseases (Gutteridge 1993; Jenner 1994; Halliwell 1996). It appears likely that certain ROS may act as complete carcinogens. Since oxidants can induce both point mutations and chromosomal aberrations in DNA, mutations in cancer-related genes including ras proto-oncogenes (Bos 1989) and *p53* tumour-suppressor genes (Hollstein *et al.* 1991) are also likely to result from ROS attack (Cerruti *et al.* 1994).

Biomarkers to study oxidative DNA damage in humans

While there are a large number of potential metabolites that can be detected to determine whole body levels of oxidative DNA damage, 8-hydroxy-deoxyguanosine (8-OHdG), 8-hydroxyguanine (8-OHG), and thymidine glycol (TG) quantitatively correlate with radiation dose (Berghold and Simic 1991) and are in general use (Lindsay 1996). One of the most used methods of determining oxidative DNA damage in intact cells and organisms is the determination of 8-OHdG using high-pressure liquid chromatography (HPLC) coupled to a highly sensitive electrochemical detector (Floyd *et al.* 1986b; Kasai and Nishimura 1984; Kasai *et al.* 1984; Auroma and Halliwell 1995). 8-Hydroxyguanine was shown to be one of the major products of ROS attack on DNA, and these lesions can be measured as to 8-OHdG by enzymatic treatment and subsequent HPLC (Auroma and Halliwell 1995). The characterisation of a number of DNA lesions, including DNA–protein complexes such as chromatin (Dizdaroglu

1991), can be measured using gas chromatography–mass spectrometry (GC–MS). DNA or chromatin is hydrolysed and the products converted to volatile derivatives, which are then separated by gas chromatography and identified by mass spectrometry (Dizdaroglu 1993). GC–MS, used in the selected ion monitoring mode (SIM), is a highly sensitive method for detecting products of DNA damage after ROS attack (Aruoma and Halliwell 1995). With respect to oxidative damage in DNA, HPLC and GC–MS both have their merits and limitations. Lipid oxidation products in plasma are best measured as isoprostanes or as lipid hydroperoxides using specific HPLC techniques (Diplock *et al.* 1998).

For *in vitro* and some *in vivo* studies, DNA damage can also be detected by identifying chemical adducts in DNA using radiolabelled carcinogens, radioimmunoassay or fluorometric techniques. An alternative approach is to measure the presence of sites sensitive to enzymes that incise DNA in regions of damage. These enzymes are called endonucleases, and they cause a reduction in the chain length of DNA. The number of sensitive sites can be determined using alkaline sucrose gradient centrifugation. Another method is to measure the reduction in the size of single-stranded DNA (ssDNA). DNA breaks can occur as a result of free radical action (at apurinic or apyriamidinic sites) or as a result of repair endonucleases. The number of single-strand breaks can be determined using alkaline sucrose gradient centrifugation, alkaline elution from membrane filters or by using the single-cell gel electrophoresis assay, more commonly known as the comet assay. The comet assay is a sensitive and rapid technique able to detect DNA strand breaks, alkali-labile sites and, by the addition of an endonuclease (*Escherichia coli* endonuclease III), whether oxidised pyrimidine bases are present which provide an indication of oxidative damage (Collins *et al.* 1993). The assay is very versatile, being able to detect damage in a wide range of mammalian cells including cultured cells, cells isolated from organs *in vivo*, and human cells obtained during intervention studies. For biomonitoring, and to assess the levels of oxidative DNA damage present in humans, the lymphocyte is commonly used, either to measure directly the level of DNA damage inflicted on the cell, or to monitor how the lymphocytes respond to an oxidative challenge with ROS *ex vivo* (Duthie *et al.* 1996; Collins *et al.* 1998).

Dietary antioxidants

The antioxidant hypothesis suggests that vitamins C and E, carotenoids (see Chapters 1 and 6) and other antioxidants occurring in fruit and vegetables afford protection against heart disease and cancer by preventing oxidative damage to lipids and to DNA. Epidemiological studies also support this hypothesis and show that antioxidant nutrients such as vitamins C and E, and β-carotene may play a beneficial role in the prevention of a number of chronic disorders (Diplock *et al.* 1998). A large case–control study in more than 2500 women in Italy associated diets rich in the micronutrients β-carotene, vitamin E and calcium with a significantly lower risk of breast cancer (Franceschi, 1997). As scientific evidence accumulates, it appears more likely that free radical damage (including oxidative stress) may play an important role in the development of cancer and other age-related diseases. The action of antioxidants can inhibit carcinogenicity (Gey 1995) by scavenging free radicals, thereby reducing the formation of carcinogens (Gower 1998; Ames 1983). Epidemiological evidence has shown an inverse correlation between the plasma levels of certain antioxidants and the incidence of cancer (Ames 1983; Gey *et al.* 1987), and together with experimental information (Potter 1997) has brought to public attention that the onset of cancer can be greatly affected by diet. Diets rich in fruit and vegetables, such as those eaten in Mediterranean countries (Hill 1997) are associated with a reduced risk of cancers of a number of sites (Steinmetz and Potter 1991). More recent interpretations of epidemiological data suggest that it is the increased consumption of vegetables especially that can delay the onset of cancer (WCRF 1997), possibly because vegetables are a rich source of ascorbate, folate, carotenoids and other essential nutrients, while fruits are generally devoid of essential nutrients with the exception of vitamin C and potassium (Davison *et al.* 1993).

The protective effect of high levels of vegetables and fruit in the diet may be due to dietary antioxidants supplementing cellular antioxidant defences; endogenous systems that are not able to cope with the increasing level of oxidative damage inflicted on the cell machinery. This hypothesis is supported by recent experimental evidence that showed pretreatment of mice with vitamins A, E or C decreased the number of mycotoxin-induced DNA adducts by 70, 80 and 90% respectively (Grosse *et al.* 1997). Intervention experiments in humans have highlighted a highly significant decrease in endogenous oxidative base damage, and an increased resistance to induced oxidative damage in lymphocyte DNA of both smokers and non-smokers after supplementation of the diet for 20 weeks with vitamin C (100 mg/day), vitamin E (280 mg/day), and β-carotene (25 mg/day) (Duthie *et al.* 1996). Increasing our consumption of dietary antioxidants by increasing our consumption of fruit and vegetables may therefore play a direct role in the prevention of cancer (Block *et al.* 1992; Block 1992), consistent with the action of these compounds to inhibit the reactions of free radicals (Simic 1991). For certain species of highly reactive free radicals, however (e.g. hydroxyl), scavenging by the addition of antioxidants to the diet is probably unrealistic because of the very high concentrations of scavenger required to compete with target molecules (Halliwell *et al.* 1995).

An alternative mechanism of action is likely to be an inhibition of the precursor events in free radical generation, effectively blocking the formation pathway. Antioxidants may also inhibit mutagenic activity by inhibiting oxidant-stimulated cell division, thereby inhibiting cancer promotion (Ames *et al.* 1993).

Our diet therefore, can be a rich source of antioxidants, and in addition to the antioxidant vitamins (vitamins E and C), there are a host of other, mainly plant-derived compounds, that possess significant antioxidant activity. These include plant polyphenolic constituents, carotenoids, and a group of compounds called flavonoids: all of which may have important functions in the prevention of chronic human diseases.

Specific antioxidants

Carotenoids

There are over 600 carotenoids that occur naturally in plants, the most important identified in human serum being; α- and β-carotene, the xanthins (β-cryptoxanthin and zeaxanthin), lycopene and leutein (Krinsky 1993; Gerster 1993). Most carotenoids are pigments that give fruits and flowers their colour and provide protection from the photochemical reactants formed during photosynthesis. The carotenoids have a characteristic structure consisting of conjugated double bonds, of which many possess vitamin A activity or are able to act as pro-vitamin A while others, including lycopene, are devoid of any vitamin A activity.

In addition to their roles as precursors of retinol and retinoids, carotenoids have distinct functions of their own in animals and humans. There is evidence, both *in vitro* and *in vivo*, that carotenoids are antioxidants or radical scavengers with a broad range of potencies (Doba *et al.* 1985; Blakely *et al.* 1988; Carbonneau *et al.* 1989), including the inhibition of lipid peroxidation (Halliwell *et al.* 1995). Carotenoids also have a range of other cellular functions, including a role in protection against phototoxicity (Davison *et al.* 1993; Krinsky 1989b; Black and Mathews-Roth 1991), which may protect against skin cancers. *In vitro*, carotenoids have been shown to protect against neoplastic transformation (β-carotene, α-carotene, canthaxanthin, α-tocopherol and lycopene) (Bertram *et al.* 1991), and inhibition of the early steps in carcinogenesis (Goswami *et al.* 1989). Epidemiological studies have highlighted a consistent association between the consumption of carotenoid-rich yellow-green fruits and vegetables and protection against cancers of a number of sites, including colon, lung, liver, stomach, breast and prostate (Davison *et al.* 1993; Peto *et al.* 1981; Wolf 1982; Ziegler 1991; Byer and Perry 1992).

How a specified carotenoid inhibits DNA damage is unknown, but because of the differences in activity of the different carotenoids it is likely that more than one mechanism of action is involved.

Beta-carotene

Beta-carotene is the most studied, but not necessarily the most important of the carotenoids. It is found in many common vegetables including carrots, spinach, kale, broccoli and in mango and papaya fruits. Beta-carotene is involved with cell–cell communication, apoptosis (see Chapter 5) and gene regulation, as well as acting as a direct antioxidant (Krinsky 1993).

It has been known for some time that β-carotene is a powerful quencher of singlet oxygen and a free radical scavenger (Foote 1968; Foote et al. 1970a,b), and it was this discovery that provided the background for its apparent protection against cancer (Gerster 1993). The antioxidant capacity of β-carotene varies with oxygen tension, and its activity is greatest at the low oxygen tension found under physiological conditions (Burton and Ingold 1984; Krinsky 1989a). In vitro work has demonstrated that β-carotene not only reacts directly with peroxyl radicals but that its effect was additive in combination with α-tocopherol. In addition, β-carotene may protect endogenous α-tocopherol against peroxyl attack (Palozza and Krinsky 1991), potentially having a dual role in protection against oxidative damage. Experimental evidence has shown that β-carotene inhibits micronuclei formation (Yager et al. 1990), which is characteristic of a number of genotoxic carcinogens (Stich et al. 1984; Stich and Dunn 1986). Micronuclei are formed either due to DNA breakage that results in a broken chromosome (the micronucleus forming from the chromosomal fragment that is not attached to a centromere), or by damage to microtubules. Epidemiological evidence suggests that β-carotene has a role in the prevention of both cardiovascular disease and cancer (Gey et al. 1987; Krinsky 1991). In a recent prospective study there was an association between low levels of β-carotene in the colon and colorectal cancer (Pappalardo et al. 1996). However, the protection afforded by β-carotene may be very dependent on serum levels. While some studies have highlighted a protective effect of serum β-carotene in lung cancer (Menkes et al. 1986; Knekt et al. 1991; Stahelin et al. 1991), a more recent study has shown high levels of β-carotene to actually enhance lung cancer development in male smokers (ATBC 1994). This could be due to the pro-oxidant activity of β-carotene which has been shown to occur at high concentrations of the carotenoid (Lawlor and O'Brien 1995).

The antioxidant capacity of the β-carotene appears to be linked to the length of the polyene backbone of the molecule, and longer-chain oxycarotenoids have been shown to be more active in quenching free radicals (Terao 1989). Lycopene, for example, has double the quenching activity of β-carotene (Di Mascio et al. 1989).

Lycopene

Lycopene is twice as effective at scavenging singlet oxygen than β-carotene, and typical human plasma levels of lycopene (0.7 μM) are also slightly higher than those of β-carotene (0.5 μM) (Di Mascio et al. 1989). Using the comet assay to determine DNA strand breakage, research conducted in our laboratory also found that lycopene inhibited hydrogen peroxide-induced oxidative DNA damage (van den Boom et al. 1997a,b) in both human colonic adenocarcinoma cells (CACO-2) and human pancreatic carcinoma cells (MiaPaCa-2). From in vivo experiments in rats, we found that the administration of tomato juice, which contained high levels of lycopene, significantly reduced the level of 1,2-dimethylhydrazine (DMH)-induced DNA damage in the colon (Hambly et al. 1997). Similar results were found in a recent human intervention study, in which DNA damage was assessed in human lymphocytes using the comet assay. After a two-week depletion period, healthy, non-smoking male volunteers received a daily dose of tomato juice containing 40 mg lycopene. This supplementation of the diet resulted in a significant decrease in endogenous levels of DNA damage, supporting the hypothesis that carotenoids exert a cancer-protective effect (Pool-Zobel et al. 1997).

Flavonoids

Flavonoids are polyphenolic compounds found in food plants, and are derived from the parent compound, flavone. They are widespread in both plants and plant products, and the estimated dietary consumption is approximately 1 g per day in the USA (Pierpoint 1986). Flavonoids are naturally present in fruits and vegetables, tea and in red wine (Kuhnau 1976), providing the food with colour, texture and taste (Harbourne 1986). Flavonoids have become a focus for nutrition research because of the anomaly known as the 'French paradox' (Formica and Regelson 1995), which refers to the lower incidence of coronary vascular disease (CVD) found in the Mediterranean population compared to other Western cultures, despite the similar fat content of the diet. It is thought that it is the flavonoids present in red wine that have the potential to reduce death from CVD (Hertog et al. 1993). Experimental data show that flavonoids are also associated with the inhibition of tumorigenesis and tumour growth in vivo (Huang et al. 1992; Chung 1992).

The flavonoids are split into major groups including the anthocyanins, flavonols, flavones, cathechins and flavanones. In addition to their biological roles in humans as important compounds involved in the maintenance of capillary walls (Harsteen 1983) the flavonoids possess a range of antioxidant defence mechanisms. These include the scavenging of lipid peroxyl radicals (Sorata et al. 1984), singlet oxygen and hydroxyl radicals (Husain et al. 1987), and superoxide anions (Robak and Gryglewski 1985), together with the sequestration of metal ions (Takahama 1985); although some can act as pro-oxidants in the presence of copper ions (Cu^{2+}).

Quercetin is one of the most common flavonoids, making up approximately 5% of the total flavonoids consumed in the diet (Brown 1980). While in several assay systems quercetin has been found to be potentially mutagenic (Bjeldans and Chong 1977; Stavric 1984; Van der Hoeven et al. 1984; Reuff et al. 1992), many long-term in vivo studies have found quercetin to be non-carcinogenic (Formica and Regelson 1995; Zhu et al. 1994; Stoewsand et al. 1984). Furthermore, quercetin was shown to possess anticarcinogenic properties (Armand et al. 1988), thought to be due to either its ability to act as an antioxidant and quench single oxygen (Chang et al. 1985), or to its ability to interact with and stabilise the secondary structure of DNA (Alvi et al. 1986).

The behaviour of quercetin to act as an antioxidant or pro-oxidant appears therefore to depend on the redox potential of the biological environment (Formica and Regelson 1995), the availability of metal ions, and the concentration of the flavonoid (Laughton et al. 1991). Further investigations of flavonoid mechanisms have highlighted the role of flavonoids in the scavenging of free radicals, and for quercetin as an inhibitor of both lipid peroxidation (Das and Ray 1988; Afanas'ev et al. 1989) and oxidation of low-density lipoproteins (de Whalley et al. 1990). The mechanism of action for quercetin was thought to be the result of electron donation to α-tocopherol (vitamin E). This sparing activity prevented the oxidation of low density lipoproteins (LDL), which in turn prevented the formation of 'foam cells' which are thought to lead to the production of atherosclerotic plaques and the subsequent development of cardiovascular disease (de Whalley et al. 1990).

Vitamins

Vitamin E (tocopherols)

There is substantial epidemiological evidence suggesting that the incidence of chronic disease is lowered in populations which have a high level of antioxidants in their diet, or who take dietary supplements such as vitamin

E (Diplock 1997; Yu *et al.* 1998). Vitamin E is an essential lipid-soluble vitamin in the human diet, and is found in cell membranes and plasma lipoproteins. Severe deprivation of vitamin E is known to lead to neurological damage (Muller and Goss-Sampson 1990). As an antioxidant active in lipid-soluble compartments of the cell, vitamin E contains a number of tocopherols. Alpha-tocopherol is one of the most important free radical scavengers and inhibits lipid peroxidation in cell membranes by scavenging peroxyl radicals. The resulting α-tocopherol radical is much less reactive and therefore slows down the peroxidation chain reaction (Burton *et al.* 1983; Burton and Traber 1990), and inhibits hydroperoxide damage to DNA (Yang and Schaich 1996). A host of other tocopherols (beta, gamma and delta) are known also to quench singlet oxygen (Sies *et al.* 1992), and there is evidence which shows that dietary vitamin E protects against oxidative DNA damage in human lymphocytes. Other evidence of a protective effect by vitamin E has been seen in its suppression of LDL oxidation both *in vitro* and *in vivo* (Yu *et al.* 1998), and evidence that vitamin E both inhibits the growth and induces apoptosis of tumour cells *in vitro* (Sigounas *et al.* 1997). Vitamin E also appears to improve the inhibition of tumour growth by the cytotoxic drug fluorouracil (5-FU), both *in vitro* and *in vivo* (Chinery *et al.* 1997). Vitamin E, as well as vitamin A, appears to inhibit bile acid induced DNA damage, and may therefore have a protective role against cancer of the colon (Booth *et al.* 1997). However, α-tocopherol has been shown to act as a potent DNA-damaging agent in the presence of Cu (II) ions *in vitro* (Yashamita *et al.* 1998), suggesting that, like most other antioxidants, vitamin E may possess pro-oxidant characteristics under particular physiological circumstances.

Vitamin C

Water-soluble ascorbic acid is essential in the diet, with a recommended daily intake of between 40 and 60 mg (Halliwell 1994b). It fulfils a number of essential metabolic functions including participating in the synthesis of hormones and neurotransmitters, as a cofactor for several enzymes including lysyl-, prolyl- and dopamine-b-hydroxylases (Levine 1986), and in the activity of cytochrome P450 enzymes (Sauberlich 1984). Its gross deficiency from the diet causes scurvy which is associated with the oxidative inactivation of certain enzymes (Meister 1994). Epidemiological evidence is very favourable for protection by vitamin C against cancer, with a majority of studies (33/46) showing a protective effect by vitamin C (or a diet high in fruit) against cancers of the oesophagus, larynx, oral cavity and pancreas (Block 1991b). In addition, vitamin C may prevent the formation of nitrosamine carcinogens in the gut (Block 1991b), thus preventing the onset of stomach cancer. Recent results from a human intervention study highlighted significant protection against oxidative DNA damage in human lymphocytes two to four hours after vitamin C intake (Panayiotidis and Collins 1997).

As an antioxidant, vitamin C is more effective in its role as a scavenger than other water-soluble antioxidants such as bilirubin or urate. Vitamin C is instrumental in scavenging a number of ROS (Bendich *et al.* 1986) including superoxide, hydroxyl radicals (Cozzi *et al.* 1997), singlet oxygen and peroxyl, together with some reactive nitrogen species including $NO_2^{\cdot -}$ (Halliwell 1994b). Vitamin C also appears to preserve the function of other antioxidants, most noticeably vitamin E, by returning them to their reduced form after quenching free radicals (Block 1991a,b). There appears to be some discrepancy in available data for the dietary effects of vitamin C. In a recent study, it was suggested that vitamin C plays an important role in the regulation of DNA repair enzymes, demonstrating a non-scavenging antioxidant role. The authors found that the levels of 8-oxo-2'-deoxyguanosine (8-oxodG) in mononuclear cell DNA, serum and urine from subjects undergoing supplementation with 500 mg/day vitamin C, were significantly decreased; correlating strongly with increased plasma vitamin C concentration (Cooke *et al.* 1998). A separate study has shown iron/ascorbate supplementation to increase biomarkers of DNA damage in well-nourished subjects (Farinati *et al.* 1998). This is not surprising, as it is well known that, in circumstances where there are high metal ion concentrations, such as after tissue injury, vitamin C can act as a pro-oxidant causing the liberation of ROS, including $OH^{\cdot}$ (Halliwell and Gutteridge 1989).

Vitamin A

Derivatives of vitamin A have been shown to inhibit tumour promotion (Verma *et al.* 1979), and both retinol and some vitamin A analogues (retinyl acetate and retinoic acid) are known to inhibit the production of superoxide in PMA-stimulated neutrophils (Witz *et al.* 1980; Kensler and Trush 1981). However, plasma vitamin A levels are under strict control and a high supplementary intake in the diet is unlikely to be relevant to cancer prevention (van Poppel and van den Berg 1997).

With such a wealth of information regarding the effect of diet on chronic disease, it would be reasonable to assume that adding specific supplements to the diet, such as vitamins or other antioxidants, would also show a protective effect. This, however, does not appear to be the case, with many recent clinical trials finding preventative antioxidant treatments to be ineffective (Dreher and Junod 1996). Some specific studies did, however, highlight some interesting results. The consumption of

Brussels sprouts, for example, appears to decrease the level of 8-OHG in the urine; implying that there was a reduction in oxidative DNA damage. Other effective regimes included a reduction in DNA damage in sperm after supplementing the diet with vitamin C, and decreased oxidative DNA damage in leukocytes upon adaption of a low fat diet (Loft and Paulsen 1996). One interesting observation was the inability of even high levels of vitamins C and E to influence the steady-state level of oxidative damage to guanine in the livers of normal unstressed guinea pigs (Cadenas et al. 1997). This suggests that these vitamins may only play a role as antioxidants in times of oxidative stress. The difficulties of interpreting antioxidant intervention may be explained by the complexities of free radical chemistry, antioxidant interactions and cancer development, but may depend also on serum levels being within narrow limits. While diet-derived antioxidants may be important in protecting against age-related diseases (cancer, CVD and neurodegenerative disease), the presence in tissues of these antioxidant vitamins may simply reflect that the individual consumes a healthy diet, and other complex mechanisms, in addition to antioxidant activity, may be responsible for protection against many of these diseases (Halliwell and Aruoma 1991). A number of dietary components such as ascorbate, some flavonoids and carotenoids can also exert pro-oxidant actions (Halliwell 1996). The physiological relevance of this is unknown, but suggests that high levels of these compounds may be deleterious, and that there is an, as yet undetermined, optimal intake of these specific antioxidants. In addition, antioxidant or pro-oxidant characteristics may vary depending on the cellular oxygen environment (Schwartz 1996).

The activity of a single antioxidant is likely to be influenced by other antioxidants present. Ascorbic acid, for example, can reduce tocopheroxyl radicals (the radical formed by the reaction of tocopherol with a peroxyl radical) regenerating tocopherol in vitro (Sies et al. 1992; McCay, 1985); thus altering the balance of antioxidant activities present in the cell. This makes it extremely difficult to discern the overall influence of one specific antioxidant in the complex environment of the cell, and even more so throughout the entire human body.

It would appear from epidemiological studies that diets high in antioxidants do have beneficial effects. Inverse relationships between the consumption of a number of single antioxidants or single vegetable supplements and cancer have been highlighted in this chapter, but other studies have shown that the greatest protective effects are seen when combinations of vegetables are consumed (Le Marchand et al. 1989). Indeed, supplementation with a single agent could elevate serum levels to such a degree that the protective, antioxidant activity is lost, and is replaced by a pro-oxidant effect. This suggests that it is not specific antioxidants in isolation that have a protective effect, but the combination of all the antioxidants present in our complex diet. Recent experimental evidence also highlights this theory. Using a micronucleus test, Konopacka and colleagues showed that pretreatment with vitamin E (100–200 mg/kg per day) or β-carotene (3–12 mg/kg per day) was effective in protecting against micronucleus induction by gamma rays. However, the most effective protection was noted when a mixture of vitamins was used (Konopacka *et al.* 1998). Similarly, recent research in our laboratory evaluated antioxidants in isolation, and plant-derived extracts with high antioxidant activity, either singly or as a complex mixture. Lipid-soluble extracts were isolated from a methanol solute of a freeze-dried vegetable sample using sodium chloride and 1:1 mixture of n-heptane and ether, and from the vegetable solid using ethylacetate. Water-soluble extracts were isolated using methanol. The total antioxidant activity of each extract was determined spectrophotometrically (Table 2.1), using the reduction of a colourless ferric iron complex to a blue complex by antioxidants present in the sample, and this was compared to the reducing ability on 1 mM Trolox (Benzie and Strain 1996).

A 'super extract' was made by combining both lipid-soluble and water-soluble fractions of the vegetable extracts (1 μM equivalents of each). Extracts were incubated with CACO-2 human colonic adenocarcinoma cells for 15 minutes prior to addition of hydrogen peroxide for 30 minutes at 37 °C to induce oxidative DNA damage. With the exception of the water-soluble extracts of tea (see Chapter 5), incubation with individual extracts alone did not induce DNA damage, nor did the incubations inhibit hydrogen-peroxide-induced DNA damage. The 'super extract' did not increase endogenous levels of DNA damage, but significantly reduced oxidative DNA damage induced by the incubation with hydrogen peroxide (Hambly 1997).

Table 2.1. Antioxidant activity of vegetable and tea extracts. Figures are mM Trolox equivalents

Extract	Lipid soluble	Water soluble
Peas	0.29	0.36
Spinach	32.0	2.56
Broccoli	8.04	1.72
Sprouts	4.3	2.91
Black tea	49.3	65.9
Green tea	7.8	177.9

While the epidemiological data for antioxidants is very promising, it should be remembered that there are a host of other compounds in fruits and vegetables in addition to antioxidants. Indeed, most antioxidants have other possibly more important functions in biological systems which may account for their protective effects.

It has been proposed that an individual at risk from a particular cancer or carcinogenic process may in the future be able to take an appropriate supplement which will act as an anticarcinogen, specific to their individual genetics or environmental risk factors (Davison *et al.* 1993). With the progress in molecular biology and epidemiology this could certainly be a feasible proposition. However, for the foreseeable future it is likely that the recommendation for protection against the diseases of increasing age will still be a diet containing a high fruit and vegetable content, as a protectant for DNA and some other cellular biomolecules.

References

Afanas'ev, I.B., Dorozhko A.I., Brodskii A.V., Kostyuk V.A. and Potapovitch A.I. (1989) Chelating and free radical scavenging mechanisms of inhibitory action of rutin and quercetin in lipid peroxidation. *Biochem. Pharmacol.*, **38**, 1763–1769.

Alvi, N.K., Rizvi R.Y. and Hadi S.M. (1986) Interaction of quercetin with DNA. *Biosci. Rep.*, **6**, 861–868.

Ames, B.N. (1983) Dietary carcinogens and anticarcinogens. Oxygen radicals and degenerative diseases. *Science*, **221**, 1256–1264.

Ames, B.N., Gold, L.S. and Willett, W.C. (1995) The causes and prevention of cancer. *Proc. Natl. Acad. Sci. US*, **92**, 5258–5265.

Ames, B.N., Shigenaga, M.K. and Hagen, T.M. (1993) Oxidants, antioxidants, and the degenerative diseases of aging. *Proc. Natl. Acad. Sci. US*, **90**, 7915–22.

Armand, J.P., de Formi M., Recondo, L., Cals, E., Cvitkovic, E. and Munch, J.N. (1988) Flavonoids: a new class of anticancer agents. Preclinical and clinical data of flavone acetic acid. In: *Progress in Clinical and Biological Research*, Vol. 280, V. Cody, E. Middleton, Jr, J.B. Harbourne and A. Beretz, eds. Alan R. Liss, New York, pp. 235–241.

Aruoma, O.I. and Halliwell, B. (1995) *DNA Damage by Free Radicals: Carcinogenic Implications, Immunopharmacology of Free Radical Species.* Academic Press, pp. 199–214.

Aruoma, O.I., Halliwell, B. and Dizdaroglu, M. (1989) Iron ion-dependent modification of bases in DNA by the superoxide radical-generating system hypoxanthine/xanthine oxidase. *J. Biol. Chem.*, **264**, 13024–13028.

ATBC (1994) The effect of vitamin E and beta carotene on the incidence of lung cancer and other cancers in male smokers. The Alpha-Tocopherol, Beta Carotene Cancer Prevention Study Group. *New Engl. J. Med.*, **330**, 1029–35.

Bartsch, H. (1996) DNA adducts in human carcinogenesis: etiological relevance and structure–activity relationship. *Mutat. Res.*, **340**, 67–79.

Beckman, J.S., Chen, J., Ischiropoulos, H. and Crow, J.P. (1994) Oxidative chemistry of peroxynitrite. *Method. Enzymol.*, **233**, 229–240.

Bendich, A., Machlin, L.J., Scandurra, O., Burton, G.W. and Wayner, D.D.M. (1986) The antioxidant role of vitamin C. *Adv. Free. Radical Biol. Med.*, **2**, 419–444.

Benzie, I.F.F. and Strain, J.J. (1996) The ferric reducing ability of plasma (FRAP) as a measure of 'antioxidant power': the FRAP assay. *Analyt. Biochem.*, **239**, 70–76.

Bergthold, D.S. and Simic, M.G. (1991) Hydroxy radical in radiation dosimetry and metabolism: dietary caloric effects. *Prog. Clin. Biol. Res.*, **372**, 21–32.

Bertram, J.S., Pung, A., Churley, M., Kappock, T.J.d., Wilkins, L.R. and Cooney, R.V. (1991) Diverse carotenoids protect against chemically induced neoplastic transformation. *Carcinogenesis*, **12**, 671–678.

Bjeldans, L.F. and Chong, G.W. (1977) Mutagenic activity of quercetin and related compounds. *Science*, **197**, 577–578.

Black, H.S. and Mathews-Roth, M.M. (1991) Protective role of butylated hydroxytoluene and certain carotenoids in photocarcinogenesis. *Photochem. Photobiol.*, **53**, 707–716.

Blakely, S.R., Slaughter, L., Adkins, J. and Knight, E.V. (1988) Effects of beta-carotene and retinyl palmitate on corn oil-induced superoxide dismutase and catalase in rats. *J. Nutr.*, **118**, 152–158.

Block, G. (1991a) Epidemiologic evidence regarding vitamin C and cancer. *Am. J. Clin. Nutr.*, **54**, 1310S–1314S.

Block, G. (1991b) Vitamin C and cancer prevention: the epidemiologic evidence. *Am. J. Clin. Nutr.*, **53**, 270S–282S.

Block, G. (1992) The data support a role for antioxidants in reducing cancer risk. *Nutr. Rev.*, **50**, 207–213.

Block, G., Patterson, B. and Subar, A. (1992) Fruit, vegetables, and cancer prevention: a review of the epidemiological evidence. *Nutr. Cancer*, **18**, 1–29.

Booth, L.A., Gilmore, I.T. and Bilton, R.F. (1997) Secondary bile acid induced DNA damage in HT29 cells: are free radicals involved? *Free Radical Res.*, **26**, 135–44.

Bos, J. (1989) ras oncogenes in human cancer: a review. *Cancer Res.*, **49**, 4682–4689.

Brawn, K. and Fridovich, I. (1981) DNA strand scission by enzymically generated oxygen radicals. *Arch. Biochem. Biophys.*, **206**, 414–419.

Brookes, P. (1977) Mutagenicity of polycyclic aromatic hydrocarbons. *Mutat. Res.*, **39**, 257–283.

Brown, J.P. (1980) A review of the genetic effects of naturally occurring flavonoids, anthroquinones and related compounds. *Mutat. Res.*, **75**, 243–277.

Burcham, P.C. (1998) Genotoxic lipid peroxidation products: their DNA damaging properties and role in formation of endogenous DNA adducts. *Mutagenesis*, **13**, 287–305.

Burdon, R.H. and Rice-Evans, C. (1989) Free radicals and the regulation of mammalian cell proliferation. *Free Radical Res. Commun.*, **6**, 345–358.

Burton, G.W. and Ingold, K.U. (1984) Beta-carotene: an unusual type of lipid antioxidant. *Science*, **224**, 569–573.

Burton, G.W. and Traber, M.G. (1990) Vitamin E: antioxidant activity, biokinetics, and bioavailability. *Ann. Rev. Nutri.*, **10**, 357–382.

Burton, G.W., Joyce A. and Ingold K.U. (1983) Is vitamin E the only lipid-soluble, chain-breaking antioxidant in human blood plasma and erythrocyte membranes? *Arch. Biochem. Biophys.*, **221**, 281–290.

Byers, T. and Perry, G. (1992) Dietary carotenes, vitamin C, and vitamin E as protective antioxidants in human cancers. *Ann. Rev. Nutr.*, **12**, 139–359.

Cadenas, S., Barja, G., Poulsen, H.E. and Loft, S. (1997) Oxidative DNA damage estimated by oxo8dG in the liver of guinea-pigs supplemented with graded dietary doses of ascorbic acid and alpha-tocopherol. *Carcinogenesis*, **18**, 2373–2377.

Carbonneau, M.A., Melin, A.M., Perromat, A. and Clerc, M. (1989) The action of free radicals on *Deinococcus radiodurans* carotenoids. *Arch. Biochem. Biophys.*, **275**, 244–251.

Cerutti, P., Ghosh R., Oya, Y. and Amstad, P. (1994) The role of cellular antioxidant defense in oxidant carcinogenesis. *Environ. Health Perspect.*, **102** (Suppl 10), 123–129.

Cerutti, P.A. (1985) Prooxidant states and tumor promotion. *Science*, **227**, 375–381.

Chance, B., Sies, H. and Boveris, A. (1979) Hydroperoxide metabolism in mammalian organs. *Physiol. Rev.*, **59**, 527–605.

Chang, R.L., Huang, M.T., Wood, A.W., Wong, C.Q., Newmark, H.L., Yagi, H., Sayer, J.M., Jerina, D.M. and Coney, A.H. (1985) Effect of ellagic acid and hydroxylated flavonoids on the tumorigenicity of benzo(a)pyrene and (+-)-7B,8a-dihydroxy-9a,10a-epoxy-7,8,9,10-tetrabenzo(a)pyrene on mouse skin and in newborn mouse. *Carcinogenesis*, **6**, 1127–1133.

Chinery, R., Brockman, J.A., Peeler, M.O., Shyr, Y., Beauchamp, R.D. and Coffey R.J. (1997) Antioxidants enhance the cytotoxicity of chemotherapeutic agents in colorectal cancer: a p53-independent induction of p21WAF1/CIP1 via C/EBPbeta [see comments]. *Nature Med.*, **3**, 1233–1241.

Chung, F.L., Xu, Y., Ho, C.T., Desai, D. and Han, C. (1992) Protection against tobacco-specific nitrosamin-induced lung tumorigenesis by green tea and its components. In: *Phenolic Compounds in Food and their Effects on Health II: Antioxidants and Cancer Prevention*, M.T. Huang, C.T. Ho and C.Y. Lee, eds. American Chemical Society, Washington, pp. 300–307.

Collins, A.R., Duthie, S.J. and Dobson, V.L. (1993) Direct enzymic detection of endogenous oxidative base damage in human lymphocyte DNA. *Carcinogenesis*, **14**, 1733–1735.

Collins, A.R., Gedik, C.M., Olmedilla, B., Southon, S. and Bellizzi, M. (1998) Oxidative DNA damage measured in human lymphocytes: large differences between sexes and between countries, and correlations with heart disease mortality rates. *FASEB J.*, **12**, 1397–1400.

Cooke, M.S., Evans, M.D., Podmore, I.D., Herbert, K.E., Mistry, N., Mistry, P., Hickenbotham, P.T., Hussieni, A., Griffiths, H.R. and Lunec, J. (1998) Novel repair action of vitamin C upon in vivo oxidative DNA damage. *FEBS Lett.*, **439**, 363–367.

Cozzi, R., Ricordy, R., Aglitti, T., Gatta, V., Perticone, P. and De Salvia, R. (1997) Ascorbic acid and beta-carotene as modulators of oxidative damage. *Nutr. Cancer*, **27**, 122–130.

Curnutte, J.T. and Babior, B.M. (1987) Chronic granulomatous disease. *Adv. Human Genet.*, **16**, 229–297.

Das, M. and Ray, P.K. (1988) Lipid antioxidant properties of quercetin in vitro. *Biochem. Int.*, **17**, 203–208.

Davison, A., Rousseau, E. and Dunn, B. (1993) Putative anticarcinogenic actions of carotenoids: nutritional implications. *Can. J. Physiol. Pharmacol.*, **71** (9), 732–745.

De Whalley, C.V., Rankin, S.M., Hoult, J.R.S., Jessup, W. and Leake, D.E. (1990) Flavonoids inhibit the oxidative modification of low density lipoproteins by macrophages. *Biochem. Pharmacol.*, **39**, 1743–1750.

Di Mascio, P., Kaiser, S. and Sies, H. (1989) Lycopene as the most efficient biological carotenoid singlet oxygen quencher. *Arch. Biochem. Biophys.*, **274**, 532–538.

Diplock, A.T. (1997) Will the 'good fairies' please prove to us that vitamin E lessens human degenerative disease? [corrected and republished in *Free Radical Res.*, 1997, Nov. 27 (5), 511–532]. *Free Radical Res.*, **26**, 565–583.

Diplock, A.T., Charleux, J.L., Crozier-Willi, G., Kok, F.J., Rice-Evans, C., Roberfroid, M., Stahl, W. and Vina-Ribes, J. (1998) Functional food science and defence against reactive oxidative species. *Br. J. Nutr.*, **80** (Suppl 1), S77–112.

Dizdaroglu, M. (1991) Chemical determination of free radical-induced damage to DNA. *Free Radical Biol. Med.*, **10**, 225–242.

Dizdaroglu, M. (1993) Quantitative determination of oxidative base damage in DNA by stable isotope-dilution mass spectrometry. *FEBS Lett.*, **315**, 1–6.

Doba, T., Burton G.W. and Ingold, K.U. (1985) Antioxidant and co-antioxidant activity of vitamin C. The effect of vitamin C, either alone or in the presence of vitamin E or a water-soluble vitamin E analogue, upon the peroxidation of aqueous multilamellar phospholipid liposomes. *Biochim. Biophys. Acta*, **835**, 298–303.

Doll, R. (1992) The lessons of life: keynote address to the nutrition and cancer conference. *Cancer Res.*, **52** (Suppl. 7), 2024s–2029s.

Doll, R. and Peto, R. (1981) The causes of cancer: quantitative estimates of avoidable risks of cancer in the United States today. *J. Natl. Cancer Inst.*, **66**, 1191–1308.

Dreher, D. and Junod A.F. (1996) Role of oxygen free radicals in cancer development. *Eur. J. Cancer.*, **32A**, 30–38.

Duthie, S.J., Ma, A., Ross, M.A. and Collins, A.R. (1996) Antioxidant supplementation decreases oxidative DNA damage in human lymphocytes. *Cancer Res.*, **56**, 1291–1295.

Fahey, R.C. (1988) Protection of DNA by thiols. *Pharmacol. Ther.*, **39**, 101–108.

Farinati, F., Cardin R., Degan P., Rugge M., Mario F.D., Bonvicini P. and Naccarato R. (1998) Oxidative DNA damage accumulation in gastric carcinogenesis. *Biochem. Biophys. Res. Commun.*, **246**, 293–298.

Fehey, R.C. and Sundquist, A.R. (1991) Evolution of glutathione metabolism. *Adv. Enzymol. Relat. Areas Mol. Biol.*, **64**, 1–53.

Floyd, R.A., Watson, J.J., Haris, J., West, M. and Wong, P.K. (1986a) Formation of 8-hydroxydeoxyguanosine, hydroxyl free radical adduct of DNA in granulocytes exposed to the tumor promoter, tetradecanoylphorbolacetate. *Biochem. Biophys. Res. Commun.*, **137**, 841–846.

Floyd, R.A., Watson, J.J., Wong, P.K., Altmiller, D.H. and Rickard, R.C. (1986b) Hydroxyl free radical adduct of deoxyguanosine: sensitive detection and mechanisms of formation. *Free Radical Res. Commun.*, **1**, 163–172.

Foote, C.S. (1968) Mechanisms of photosensitized oxidation. There are several different types of photosensitized oxidation which may be important in biological systems. *Science*, **162**, 963–970.

Foote, C.S., Chang, Y.C. and Denny, R.W. (1970a) Chemistry of singlet oxygen. X. Carotenoid quenching parallels biological protection. *J. Am. Chem. Soc.*, **92**, 5216–5218.

Foote, C.S., Chang, Y.C. and Denny, R.W. (1970b) Chemistry of singlet oxygen. XI. *Cis-trans* isomerization of carotenoids by singlet oxygen and a probable quenching mechanism. *J. Am. Chem. Soc.*, **92**, 5218-5219.

Formica, J.V. and Regelson W. (1995) Review of the biology of quercetin and related bioflavonoids. *Food Chem. Toxicol.*, **33**, 1061–1080.

Fraga, C.G., Shigenaga, M.K., Park, J.W., Degan, P. and Ames, B.W. (1990) Oxidative damage to DNA during aging: 8-hydroxy-2'-deoxyguanosine in rat organ DNA and urine. *Proc. Natl. Acad. Sci. US*, **87**, 4533–4537.

Franceschi, S. (1997) Micronutrients and breast cancer. *Eur. J. Cancer Prev.*, **6**, 535–539.

Fridovich, I. (1978) The biology of oxygen radicals. The superoxide radical is an agent of oxygen toxicity; superoxide dismutase provides an important defence. *Science*, **201**, 875–880.

Fridovich, I. (1983) Superoxide radical: an endogenous toxicant. *Ann. Rev. Pharmacol. Toxicol.*, **23**, 239–257.

Fridovich, I. (1986) Biological effects of the superoxide radical. *Arch. Biochem. Biophys.*, **247**, 1–11.

Fujiki, H., Mori, M., Nakayasu, M., Tereda, T., Sugimura, T. and Moore, R.E. (1981) Indole alkaloids: dihydroteleocidin B, telocidin, and lyngbyatoxin A as members of a new class of tumor promoters. *Proc. Natl. Acad. Sci. US*, **78**, 3872–3876.

Gerster, H. (1993) Anticarcinogenic effect of common carotenoids. *Int. J. Vitam. Nutr. Res.*, **63**, 93–121.

Gey, F., Brubacher, G.B. and Stahelin, H.B. (1987) Plasma levels of antioxidant vitamins in relation to ischemic heart disease and cancer. *Am. J. Clin. Nutr.*, **45**, 1368–1377.

Gey, K.F. (1995) 10-Year retrospective on the antioxidant hypothesis of atherosclerosis – threshold plasma levels of antioxidant micronutrients related to minimum cardiovascular risk. *J. Nutr. Biochem.*, **6**, 206–236.

Goswami, U.C., Saloi T.N., Firozi P.F. and Bhattacharya R.K. (1989) Modulation by some natural carotenoids of DNA adduct formation by aflatoxin B1 in vitro. *Cancer Lett.*, **47**, 127–132.

Gower, J.D. (1988) A role for dietary lipids and antioxidants in the activation of carcinogens. *Free Radical Biol. Med.*, **5**, 95–111.

Greenstock, C.L. (1993) Radiation and aging: free radical damage, biological response and possible antioxidant intervention. *Med. Hypoth.*, **41**, 473–482.

Grosse, Y., Chekir-Ghedira, L., Huc, A., Obrecht-Pflumio, S., Dirheimer, G., Bacha, H. and Pfohl-Leszkowicz, A. (1997) Retinol, ascorbic acid and alpha-tocopherol prevent DNA adduct formation in mice treated with the mycotoxins ochratoxin A and zearalenone. *Cancer Lett.*, **114**, 225–229.

Gutteridge, J.M.C. (1993) Free radicals in disease processes: a compilation of cause and consequence. *Free Radical Res. Commun.*, **19**, 141–158.

Halliwell, B. (1978a) Superoxide-dependent formation of hydroxyl radicals in the presence of iron chelates: is it a mechanism for hydroxyl radical production in biochemical systems? *FEBS Lett.*, **92**, 321–326.

Halliwell, B. (1978b) Superoxide-dependent formation of hydroxyl radicals in the presence of iron salts. Its role in degradation of hyaluronic acid by a superoxide-generating system. *FEBS Lett.*, **96**, 238–242.

Halliwell, B. (1987) Oxidants and human disease: some new concepts. *FASEB J.*, **1**, 358–364.

Halliwell, B. (1994a) Free radicals and antioxidants: a personal view. *Nutr. Rev.*, **52**, 253–265.

Halliwell, B. (1994b) Vitamin C: the key to health or a slow-acting carcinogen. *Redox Rep.*, **1**, 5–9.

Halliwell, B. (1995) Oxidation of low-density lipoproteins: questions of initiation, propagation, and the effect of antioxidants. *Am. J. Clin. Nutr.*, **61** (Suppl.), 670s–677s.

Halliwell, B. (1996) Oxidative stress, nutrition and health. Experimental strategies for optimization of nutritional antioxidant intake in humans. *Free Radical Res.*, **25**, 57–74.

Halliwell, B. and Aruoma, O.I. (1991) DNA damage by oxygen-derived species. Its mechanism and measurement in mammalian systems. *FEBS Lett.*, **281**, 9–19.

Halliwell, B. and Gutteridge, J.M. (1984) Free radicals, lipid peroxidation, and cell damage [letter]. *Lancet*, **2**, 1095.

Halliwell, B. and Gutteridge, J.M.C. (1989) *Free Radicals in Biology and Medicine*, 2nd edn. Clarendon Press, Oxford.

Halliwell, B., Aeschbach, R., Loliger, J. and Aruoma, O.I. (1995) The characterization of antioxidants. *Food Chem. Toxicol.*, **33**, 601–617.

Hambly, R.J. (1997) Antigenotoxic activity of vegetable extracts. Rank Prize Fund Symposium, Windermere, Lake District.

Hambly, R.J., Malcolm, C. and Rowland, I.R. (1997) Inhibition of DNA damage by lycopene in vivo. Japanese Society of Cancer Prevention, Japan.

Harbourne, J.B. (1986) Nature, distribution and function of plant flavonoids. In: *Progress in Clinical and Biological Research*, Vol 213, V. Cody, E. Middleton Jr and J.B. Harbourne, eds. Alan R. Liss, New York, pp. 15–24.

Havsteen, B. (1983) Flavonoids, a class of products of high pharmacological potency. *Biochem. Pharmacol.*, **32**, 1141–1448.

Hertog, M.G., Feskens, E.J.M., Hollman, P.C.H., Katan, M.B. and Kromhout, D. (1993) Dietary antioxidant flavonoids and risk of coronary heart disease: the Zutphen Elderly Study. *Lancet*, **342**, 1007–1011.

Hill, M. (1997) Changing pattern of diet in Europe. *Eur. J. Cancer Prev.*, **6** (Suppl. 1), 11s–13s.

Hollstein, M., Sidransky D., Vogelstein B. and Harris C. (1991) p53 mutations in human cancers. *Science*, **253**, 49–53.

Huang, M.T., Wang Z.Y., Georgiadis C.A., Laskin J.D. and Conney A.H. (1992) Inhibitory effects of curcumin on tumor initiation by benzo[a]pyrene and 7,12-dimethylbenz[a]anthracene. *Carcinogenesis*, **13**, 2183–2186.

Huie, R.E. and Padmaja S. (1993) The reaction of NO with superoxide. *Free Radical Res. Commun.*, **18**, 195–199.

Husain, S.R., Cillard, J. and Cillard, P. (1987) Hydroxy radical scavenging activity of flavonoids. *Phytochemistry*, **26**, 2489–2491.

Inouye, S. (1984) Site-specific cleavage of double-strand DNA by hydroperoxide of linoleic acid. *FEBS Lett.*, **172**, 231–234.

Jenner, P. (1994) Oxidative damage in neurodegenerative disease. *Lancet*, **344**, 796–798.

Kasai, H. and Nishimura, S. (1984) Hydroxylation of deoxyguanosine at the C-8 position by ascorbic acid and other reducing agents. *Nucleic Acids Res.*, **12**, 2137–2145.

Kasai, H., Crain, P.F., Kuchino, Y., Nashimura, S., Ootsuyama, A. and Tanooka, H. (1986) Formation of 8-hydroxyguanine moiety in cellular DNA by agents producing oxygen radicals and evidence for its repair. *Carcinogenesis*, **7**, 1849–1851.

Kasai, H., Tanooka, H. and Nishimura, S. (1984) Formation of 8-hydroxyguanine residues in DNA by X-irradiation. *Gann*, **75**, 1037–1039.

Kehrer, J.P. (1993) Free radicals as mediators of tissue injury and disease. *Crit. Rev. Toxicol.*, **23**, 21–48.

Kensler, T.W. and Trush, M.A. (1981) Inhibition of phorbol ester-stimulated chemiluminescence in human polymorphonuclear leukocytes by retinoic acid and 5,6-epoxyretinoic acid. *Cancer Res.*, **41**, 216–222.

Knekt, P., Jarvinen, R., Seppanen, R., Rissanen, A., Aromaa, A., Heinonen, O.P., Albanes, D., Heinonen, M., Pukkala, E. and Teppo, L. (1991) Dietary antioxidants and the risk of lung cancer [see comments]. *Am. J. Epidemiol.*, **134**, 471–479.

Konopacka, M., Widel, M. and Rzeszowska-Wolny, J. (1998) Modifying effect of vitamins C, E and beta-carotene against gamma-ray-induced DNA damage in mouse cells. *Mutat. Res.*, **417**, 85–94.

Krinsky, N.I. (1989a) Antioxidant functions of carotenoids. *Free Radical Biol. Med.*, **7**, 617–635.

Krinsky, N.I. (1989b) Carotenoids as chemopreventive agents. *Prev. Med.*, **18**, 592–602.

Krinsky, N.I. (1991) Effects of carotenoids in cellular and animal systems. *Am. J. Clin. Nutr.*, **53**, 238s–246s.

Krinsky, N.I. (1993) Actions of carotenoids in biological systems. *Ann. Rev. Nutr.*, **13**, 561–587.

Kuhnau, J. (1976) The flavonoids. A class of semi-essential food components: their role in human nutrition. *World Rev. Nutr. Diet*, **24**, 117–191.

Laughton, M.J., Evans, P.A., Moroney, M.A., Hoult, J.R.S. and Halliwell, B. (1991) Inhibition of mammalian 5-lipoxygenase and cyclooxygenase by flavonoids and phenolic dietary additives. *Biochem. Pharmacol.*, **42**, 1673–1681.

Lawlor, S.M. and O'Brien N.M. (1995) Modulation of oxidative stress by beta-carotenee in chick embryo fibroblasts. *Br. J. Nutr.*, **73**, 841–850.

Le Marchand, L., Yoshizawa, C.N., Kolonel, L.N., Hankin, J.H. and Goodman, M.T. (1989) Vegetable consumption and lung cancer risk: a population-based case–control study in Hawaii. *J. Natl. Cancer Inst.*, **81**, 1158–1164.

Levine, M. (1986) New concepts in the biology and biochemistry of ascorbic acid. *New Engl. J. Med.*, **314**, 892–902.

Lindsay, D.G. (1996) Dietary contribution to genotoxic risk and its control. *Food Chem. Toxicol.*, **34**, 423–431.

Loft, S. and Poulsen H.E. (1996) Cancer risk and oxidative DNA damage in man. *J. Mol. Med.*, **74**, 297–312.

McCay, P.B. (1985) Vitamin E: Interactions with free radicals and ascorbate. *Ann. Rev. Nutr.*, **5**, 323–340.

McCord, J.M. and Day, E.D.J. (1978) Superoxide-dependent production of hydroxyl radical catalyzed by iron–EDTA complex. *FEBS Lett.*, **86**, 139–142.

McCord, J.M., Wong, K., Stokes, S.H., Petrone, W.F. and English, D. (1982) In: *Pathology of Oxygen*, A.P. Autor, ed. Academic Press, New York, pp. 75–81.

McLennan, G., Oberley L.W. and Autor A.P. (1980) The role of oxygen-derived free radicals in radiation-induced damage and death of nondividing eucaryotic cells. *Radiat. Res.*, **84**, 122–132.

Meier, B., Radeke, H., Selle, S. *et al.* (1990) Human fibroblasts release reactive oxygen species in response to treatment with synovial fluids from patients suffering from arthritis. *Free Radical Res. Commun.*, **8**, 149–160.

Meister, A. (1994) Glutathione, ascorbate, and cellular protection. *Cancer Res.*, **54** (Suppl. 7), 1969s–1975s.

Mello Filho, A.C. and Meneghini, R. (1984) In vivo formation of single-strand breaks in DNA by hydrogen peroxide is mediated by the Haber–Weiss reaction. *Biochim. Biophys. Acta*, **781**, 56–63.

Mello Filho, A.C., Hoffmann, M.E. and Meneghini, R. (1984) Cell killing and DNA damage by hydrogen peroxide are mediated by intracellular iron. *Biochem. J.*, **218**, 273–275.

Menkes, M., Comstock, G., Vuilleumier, J., Helsing, K., Rider, A. and Brookmeyer, R. (1986) Serum-carotene, vitamins A and E, selenium, and the risk of lung cancer. *New Engl. J. Med.*, **315**, 1250–1254.

Miller, E.C. and Miller, J.A. (1981) Mechanisms of chemical carcinogenesis. *Cancer*, **47** (Suppl. 5), 1055–1064.

Muller, D.P.R. and Goss-Sampson, M.A. (1990) Neurochemical, neurophysiological, and neuropathological studies in vitamin E deficiency. *Crit. Rev. Neurobiol.*, **5**, 239–263.

Palozza, P. and Krinsky, N.I. (1991) The inhibition of radical-initiated peroxidation of microsomal lipids by both alpha-tocopherol and beta-carotene. *Free Radical Biol. Med.*, **11**, 407–414.

Panayiotidis, M. and Collins, A.R. (1997) Ex vivo assessment of lymphocyte antioxidant status using the comet assay. *Free Radical Res.*, **27**, 533–537.

Pappalardo, G., Guadalaxara, A., Maiani, G., Illomei, G., Trifero, M., Frattaroli, F.M. and Mobarhan, S. (1996) Antioxidant agents and colorectal carcinogenesis: role of beta-carotene, vitamin E and vitamin C. *Tumori*, **82**, 6–11.

Peto, R., Doll, R., Buckley, J.D. and Sporn, M.B. (1981) Can dietary beta-carotene materially reduce human cancer rates? *Nature*, **290**, 201–208.

Pierpoint, W.S. (1986) Flavonoids in the human diet. In: *Progress in Clinical and Biolological Research*, Vol. 213, V. Cody, E. Middleton Jr and J.B. Harbourne, eds. Alan R. Liss, New York, pp. 125–140.

Pool-Zobel, B.L., Bub, A., Muller, H., Wollowski, I. and Rechkemmer, G. (1997) Consumption of vegetables reduces genetic damage in humans: first results of a human intervention trial with carotenoid-rich foods. *Carcinogenesis*, **18**, 1847–1850.

Potter, J.D. (1997) Cancer prevention; epidemiology and experiment. *Cancer Lett.*, **114**, 7–9.

Poulsen, H.E., Prieme, H. and Loft, S. (1998) Role of oxidative DNA damage in cancer initiation and promotion. *Eur. J. Cancer Prev.*, **7**, 9–16.

Prutz, W.A., Butler, J. and Land, E.J. (1990) Interaction of copper(I) with nucleic acids. *Int. J. Radiat. Biol.*, **58**, 215–234.

Radi, R., Beckman, J.S., Bush, K.K.M. and Freeman, B.A. (1991) Peroxynitrite oxidation of sulphydryls. The cytotoxic potential of superoxide and nitric oxide. *J. Biol. Chem.*, **266**, 4244–4250.

Reuff, J., Laires, A., Gaspar, J., Borda, M. and Rodriguez, A. (1992) Oxygen species and the genotoxicity of quercetin. *Mutat. Res.*, **265**, 75–81.

Richter, C. (1995) Oxidative damage to mitochondrial DNA and its relationship to ageing. *Int. J. Biochem. Cell Biol.*, **27**, 647–653.

Robak, J. and Gryglewski, R.J. (1988) Flavonoids are scavengers of superoxide anions. *Biochem. Pharmacol.*, **37**, 837–841.

Rowley, D.A. and Halliwell, B. (1983) DNA damage by superoxide-generating systems in relation to the mechanism of action of the anti-tumour antibiotic adriamycin. *Biochim. Biophys. Acta.*, **761**, 86–93.

Sauberlich, H.E. (1984) Ascorbic acid. In Olsen, R.E., Broquist, H.P., Chichester, C.O., Darby, W.J., Kolbye, A.C. and Stalvey, R.M. (eds), *Nutrition Reviews' Present Knowledge in Nutrition*. The Nutrition Foundation, Washington, DC, pp. 260–272.

Schwartz, J.L. (1996) The dual roles of nutrients as antioxidants and prooxidants: their effects on tumor cell growth. *J. Nutr.*, **126**, 1221S–1227S.

Shacter, E., Beecham, E.J., Covey, J.M., John, K.W. and Potter, M. (1988) Activated neutrophils induce prolonged DNA damage in neighboring cells. *Carcinogenesis*, **9**, 2297–2304.

Sies, H., Sundquist, A.R. and Stahl, W. (1992) Antioxidant functions of vitamins. Vitamin E and C, beta-carotene, and other carotenoids. *Annu. New York Acad. Sci.*, **669**, 7–20.

Sigounas, G., Anagnostou, A. and Steiner, M. (1997) dl-alpha-tocopherol induces apoptosis in erythroleukemia, prostate, and breast cancer cells. *Nutr. Cancer*, **28**, 30–35.

Simic, M.G. (1991) DNA damage, environmental toxicants, and rate of aging. *J. Environ. Sci. Health*, **C9**, 113–153.

Singer, B. (1996) DNA damage: chemistry, repair, and mutagenic potential. *Regulat. Toxicol. Pharmacol.*, **23**, 2–13.

Sorata, Y., Takahama, U. and Kimura, M. (1984) Protective effect of quercetin and rutin on photosensitized lysis of human erythrocytes in the presence of hematoporphyrin. *Biochim. Biophys. Acta*, **799**, 313–317.

Stahelin, H.B., Gey, K.F., Eichholzer, M. and Ludin, E. (1991) Beta-carotene and cancer prevention: the Basel Study. *Am. J. Clin. Nutr.*, **53**, 265S–269S.

Stavric, B. (1984) Mutagenic food flavonoids. *Federation Proc.*, **43**, 2454–2458.

Steinmetz, K.A. and Potter, J.D. (1991) Vegetables, fruit, and cancer. I. Epidemiology. *Cancer Causes and Control*, **2**, 325–357.

Stich, H.F. and Dunn, B.P. (1986) Relationship between cellular levels of beta-carotene and sensitivity to genotoxic agents. *Int. J. Cancer.*, **38**, 713–737.

Stich, H.F., Stich, W., Rosin, M.P. and Vallejera, M.O. (1984) Use of the micronucleus test to monitor the effect of vitamin A, beta-carotene and canthaxanthin on the buccal mucosa of betel nut/tobacco chewers. *Int. J. Cancer.*, **34**, 745–750.

Stoewsand, G.S., Anderson, J.L. and Boyd, J.N. (1984) Quercetin: a mutagen, not a carcinogen in Fischer rats. *J. Toxicol. Environ. Health*, **14**, 105–114.

Takahama, U. (1985) Inhibition of dependant lipid peroxidation by quercetin – mechanism of antioxidative function. *Phytochemistry*, **24**, 1443–1446.

Terao, J. (1989) Antioxidant activity of beta-carotene-related carotenoids in solution. *Lipids*, **24**, 659–661.

Tkeshelashvili, L.T., McBride, T., Spence, K. and Loeb, L.A. (1991) Mutation spectrum of copper-induced DNA damage. *J. Biol. Chem.*, **266**, 6401–6406.

Troll, W. and Wiesner, R. (1985) The role of oxygen radicals as a possible mechanism of tumor promotion. *Ann. Rev. Pharmacol. Toxicol.*, **25**, 509–528.

Troll, W., Witz, G., Goldstein, B., Stone, D. and Sugimura, T. (1982) The role of free oxygen radicals in tumor promotion and carcinogenesis. In: *Carcinogenesis. A Comprehensive Survey. Cocarcinogenesis and Biological Effects of Tumor Promoters*, Vol. 7, E. Hecker, W. Fusenig, W. Kunz, F. Marks and H.W. Thielmann, eds. Raven, New York, pp. 593–597.

Ueda, K., Kobayashi, S., Morita, J. and Komano, T. (1985) Site-specific DNA damage caused by lipid peroxidation products. *Biochim. Biophys. Acta.*, **824**, 341–348.

Van der Hoeven, J.C., Bruggeman, I.M. and Debets, F.M. (1984) Genotoxicity of quercetin in cultured mammalian cells. *Mutat. Res.*, **136**, 9–21.

van den Boom, C., Hambly, R.J., Nozawa, H., Sonobe, H. and Rowland, I.R. (1997a) Antigenotoxic effects of lycopene against oxidative alkylating DNA damage in CACO-2 and MiaPaCa-2 cells. Carotenoids: Occurrence, food processing and physiological significance, 2nd Karlsruhe Nutrition Symposium.

van den Boom, C., Hambly, R.J., Nozawa, H., Sonobe, H. and Rowland, I.R. (1997b) Antioxidant effects of tomato juice and lycopene in CACO-2 cells. Satellite Meeting: Genotoxicity and Diet Cancer Prevention, Pisa, Italy.

van Poppel, G. and van den Berg H. (1997) Vitamins and cancer. *Cancer Lett.*, **114**, 195–202.

Verma, A.K., Shapas, B.G., Rice, H.M. and Boutwell, R.K. (1979) Correlation of the inhibition by retinoids of tumor promoter-induced mouse epidermal ornithine decarboxylase activity and of skin tumor promotion. *Cancer Res.*, **39**, 419–425.

Von Sonntag, C. (1987) *The Chemical Basis of Radiation Biology*. Taylor & Francis, London.

WCRF (1997) Food, nutrition and the prevention of cancer: a global perspective. World Cancer Research Fund and the American Institute for Cancer Research.

Wei, Q., Matanoski, G.M., Farmer, E.R., Hedayati, A.M. and Grossman, L. (1993) DNA repair and ageing in basal cell carcinoma: a molecular epidemiology study. *Proc. Natl. Acad. Sci. US*, **90**, 1614–1618.

Weitzman, S.A. and Gordon, L.I. (1990) Inflammation and cancer: role of phagocyte-generated oxidants in carcinogenesis. *Blood*, **76**, 655–663.

Witz, G., Goldstein, B.A., Amoruso, M., D.S., S. and Troll, W. (1980) Retinoid inhibition of superoxide anion radical production by human polymorphonuclear leukocytes stimulated with tumor promoters. *Biochem. Biophys. Res. Commun.*, **97**, 883–888.

Witztum, J.L. (1994) The oxidation hypothesis of atherosclerosis. *Lancet*, **344**, 793–795.

Wolf, G. (1982) Is dietary beta-carotene an anti-cancer agent? *Nutr. Rev.*, **40**, 257–261.

Yager, J.W., Eastmond, D.A., Robertson, M.L., Paradisin, W.M. and Smith, M.T. (1990) Characterization of micronuclei induced in human lymphocytes by benzene metabolites. *Cancer Res.*, **50**, 393–399.

Yamashita, N., Murata, M., Inoue, S., Burkitt, M.J., Milne, L. and Kawanishi, S. (1998) Alpha-tocopherol induces oxidative damage to DNA in the presence of copper(II) ions. *Chem. Res. Toxicol.*, **11**, 855–862.

Yang, M.H. and Schaich, K.M. (1996) Factors affecting DNA damage caused by lipid hydroperoxides and aldehydes. *Free Radical Biol. Med.*, **20**, 225–236.

Yu, B.P., Kang, C.M., Han, J.S. and Kim D.S. (1998) Can antioxidant supplementation slow the aging process? *Biofactors*, **7**, 93–101.

Zhu, B.T., Ezell E.L. and Liehr J.G. (1994) Catechol-O-methyl transferase catalysed rapid O-methylation of mutagenic flavonoids. Metabolic inactivation as a possible reason for their lack of carcinogenicity in vivo. *J. Biol. Chem.*, **269**, 292–299.

Ziegler, R.G. (1991) Vegetables, fruits, and carotenoids and the risk of cancer. *Am. J. Clin. Nutr.*, **53**, 251S–259S.

Zimmerman, R. and Cerutti, P. (1984) Active oxygen acts as a promoter of transformation in mouse embryo C3H/10T1/2/C18 fibroblasts. *Proc. Natl. Acad. Sci. US*, **81**, 2085–2087.

3 Protein Injury: Prevention by Antioxidants

BARBARA R. EVANS and JONATHAN WOODWARD
Chemical Technology Division, Oak Ridge National Laboratory[1] Oak Ridge, TN 37831-6194

Oxidative damage to proteins: causes and mechanism

Introduction

The use of antioxidants to prevent damage to proteins is familiar from its mention in sources, ranging from beginning biochemistry textbooks to the latest scientific journals. Directly or indirectly, oxidative damage is generally caused by the action of oxygen. Many of the compounds required for metabolism in modern organisms are sensitive to this type of damage. Yet most organisms one encounters today require oxygen for life, or are at least able to survive in an atmosphere containing oxygen at the current level of 20.95% by volume. The reasons for this seeming paradox may lie in the origins of life in the early history of the Earth. During the era when life evolved on Earth, the terrestrial atmosphere was thought to have been reducing, being composed of carbon dioxide, methane, ammonia and hydrogen. Based on geological evidence, estimates of the oxygen levels on the early Earth range from 0.01 to 0.2% of the modern levels (Urey 1952; Miller and Orgel 1974). According to a newer model, the main components of the early terrestrial atmosphere were carbon dioxide, water and nitrogen (Mattioli and Wood 1986; Holland 1984). The first micro-organisms may have been anaerobic methanogens similar to members of the Archaea existing today in anaerobic environments. Such micro-organisms use enzymes containing nickel and iron complexed with sulphur to produce methane and energy from hydrogen and carbon dioxide (Huber and Wächterhäuser 1997; Hahn and Kück 1994). Microbial iron reduction on the primitive Earth may have been carried out by micro-organisms similar to

[1]Managed by UT-Batelle, LLC, for the US Department of Energy under contract DE-AC05-00OR22725. The submitted manuscript has been authorised by a contractor of the US Government under contract No. DE-AC05-00OR22725. Accordingly, the US Government retains a nonexclusive, royalty-free license to publish or reproduce the published form of this contribution, or allow others to do so, for US Government purposes. Published by John Wiley & Sons Ltd, 2000.

anaerobic, deep subsurface bacteria that use hydrogen as a substrate (Liu et al. 1997). According to the fossil record, photosynthetic micro-organisms first appeared about 3 billion years ago. Fossils of eukaryotic and presumably aerobic organisms are estimated to be 1 billion years old. With the advent of photosynthesis, large amounts of oxygen began to enter the terrestrial atmosphere. Currently, oxygen is produced from photosynthesis at an average rate of about $0.075 \, \text{g/cm}$ per year, of which $3.4 \times 10^{17} \, \text{g/year}$ is produced from the ocean. Both the production of oxygen from photosynthesis and the escape of hydrogen have increased the oxidizing capacity of the atmosphere. Oxygen now composes approximately 20% of the terrestrial atmosphere. Since the average aqueous solubility of oxygen is about $10 \, \text{mg/l}$, dissolved oxygen is present in most aqueous environments at concentrations of around 0.5 to $1 \, \text{m}M$. The presence of oxygen necessitated the introduction of antioxidants and oxidation-resistant proteins and other biomolecules by life forms, as well as compartmentalisation to exclude oxygen (Miller and Orgel 1974).

Today a vast variety of organisms populate different environments on the planet, each of which offers its own challenges for survival. Some of the micro-organisms that are of great scientific and commercial interest grow under anaerobic, thermophilic conditions ranging from compost heaps to submarine thermal vents. However, a protein may not necessarily be more sensitive to oxidation because of its source. Even in the case of strict anaerobes, a wide range of oxygen sensitivity may be observed among the proteins produced by a single species. Thus, the type of antioxidant employed in a particular case depends on the type of protein under consideration and the intended application, as well as the source of that protein. Often extracellular and periplasmic enzymes from anaerobic thermophiles are very stable in aerobic conditions and can be purified, assayed and stored without stringent oxygen exclusion. Such enzymes can be readily expressed as recombinant proteins in micro-organisms such as *Escherichia coli*. Some examples are β-xylosidase (Shao and Wiegel 1992), β-galactosidase (Fokhina and Velikodvorskaia 1997), and xylose (glucose) isomerase (Dekker *et al.* 1991; Lee and Zeikus 1992) from *Thermoanaerobacter ethanolicus* (formerly *Clostridium thermohydrosulfuricum*); the two endo-1,4-β-xylanases from *Thermotoga thermarum* (Sunna *et al.* 1996); and the β-glucosidase (Kengen *et al.* 1993; Voorhorst *et al.* 1995) and the esterase (Ikeda and Clark, 1998) from the hyperthermophilic archaeon, *Pyrococcus furiosus*.

When recombinant proteins from an anaerobic micro-organism are expressed aerobically in *E. coli*, the strict exclusion of oxygen employed for purification from the source organism may be unnecessary. An example is the secondary alcohol dehydrogenase of *Thermoanaerobacter ethanolicus*. When the enzyme was purified from *T. ethanolicus*, strict anaerobic conditions were used, including carrying out purification steps in a glove bag under an atmosphere of $19:1$ $N_2:H_2$ and inclusion of $2\,mM$ dithiothreitol in all of the buffers (Burdette and Zeikus 1994). When the same enzyme was cloned and expressed in *E. coli*, the only precaution taken during its purification was the inclusion of $5\,mM$ dithiothreitol in all of the buffers used for the preparation (Burdette *et al.* 1997). However, oxidation-sensitive enzymes require inclusion of antioxidants and other stabilizers during purification regardless of the micro-organism in which the protein is expressed. Redox enzymes, particularly those composed of multiple heterologous subunits such as the hydrogenase from the *P. furiosus*, generally must be purified from the source micro-organism with the employment of stringent oxygen exclusion and the maintenance of reducing conditions. In the purification of ferrodoxin and hydrogenase from *P. furiosus*, all buffers must be degassed, followed by flushing with argon, as well as addition of $2\,mM$ sodium dithionite, to protect against any traces of oxygen still remaining (Bryant and Adams 1989; Aono *et al.* 1989; Adams 1990; Adams and Zhou 1995).

Types of oxidative damage

The reactive oxygen species (ROS) that are reported to damage proteins are hydrogen peroxide, hydroxyl radicals and superoxide radicals, all strong oxidisers (Table 3.1). The amino acid side chains that are potentially targeted by ROS include: sulphydryl groups of cysteine, amines of lysine,

Table 3.1. Selected biochemical redox potentials at pH 7 and 20 °C

System	E^0 (V)
Hydrogen peroxide/water	1.77[a]
Hydroxyl radical ($^\bullet$OH)	2.02[b]
Superoxide radical ($O_2^{\bullet-}$)	0.95 −0.33[b]
p-Quinone/hydroquinone	0.28[a]
NAD$^+$/NADH	−0.320[a]
Dehydroascorbic acid/ascorbic acid	0.058[a]
Cysteine/cystine	−0.340[a]
Glutathione reduced/oxidized	−0.340[a]
1,4-Dithiothreitol	−0.332[c]
Pentaammine Ru^{2+}/Ru^{3+}	0.090[d]
Microperoxidase-8 Fe^{3+}/Fe^{2+}	−0.201[e]

[a]Loach 1968.
[b]Guissani *et al.* 1982.
[c]Cleland 1964.
[d]Evans *et al.* 1993.
[e]Harbury and Loach 1960.

hydroxyl groups of serine and threonine and the aromatic rings of phenylalanine, tyrosine, histidine and tryptophan (Figure 3.1). In terms of protein stability, the most oxidation-sensitive moieties are the thiol of cysteine and the indole of tryptophan. Analysis of some thermophilic enzymes indicates that removal of these two amino acids by site-directed mutagenesis may increase protein stability. Alternatively, inactivation may be avoided by addition of appropriate antioxidants, substrates, and cofactors or by exclusion of oxygen (Mozhaev *et al.* 1988).

The problems of oxidative damage to recombinant proteins in aerobic culture during scaled-up production have recently been reviewed (Konz *et al.* 1998). The aerobic nature of the commonly used hosts such as *E. coli* requires a minimum oxygen level to maintain growth and maximise protein production. At the same time, oxidative damage to recombinant proteins can reduce the yield and value of the produced proteins.

Sulphydryl groups

Perhaps the most familiar oxidation-sensitive side chain of proteins is the sulphydryl group of cysteine. Oxidation of cysteine can result in formation of cysteic acid, while oxidation of two cysteines causes the formation of cystine, with the two cysteines linked through a disulphide bond. Since disulphide bonds help hold proteins in their native conformations, reduction of cystine linkages and their subsequent oxidation to form inappropriate disulphide bonds can result in structural distortion and consequently loss of functional activity. Proper refolding is dependent on protein concentration, pH and oxidation rate (Anfinsen and Haber 1961; White 1961; Fischer *et al.* 1992).

Cysteine side chains can also have catalytic functions, as in the case of the sulphydryl proteases such as papain (Balls *et al.* 1937) and cathepsins (Boyer 1960). Cysteinyl thiols act as ligands in the zinc-binding domains of proteins such as protein kinase C (Hubbard 1995) and coordinate metals in the active sites of certain metalloenzymes (Adams 1990; Peters *et al.* 1998). The small protein thioredoxin uses the oxidation of two thiol groups to a disulphide in its active site to act as a hydrogen donor for ribonucleotide reductase (Holmgren and Björnstedt 1995). Cysteinyl thiol groups perform important catalytic functions in photosynthetic systems as well. Reagents that specifically modify sulphydryl groups abolish photophosphorylation by isolated spinach chloroplast thylakoids (Moroney *et al.* 1980).

Lysine

The ε-amine of lysine and the free amino group of the amino terminal of proteins can form crosslinks through reaction with carbonyl groups

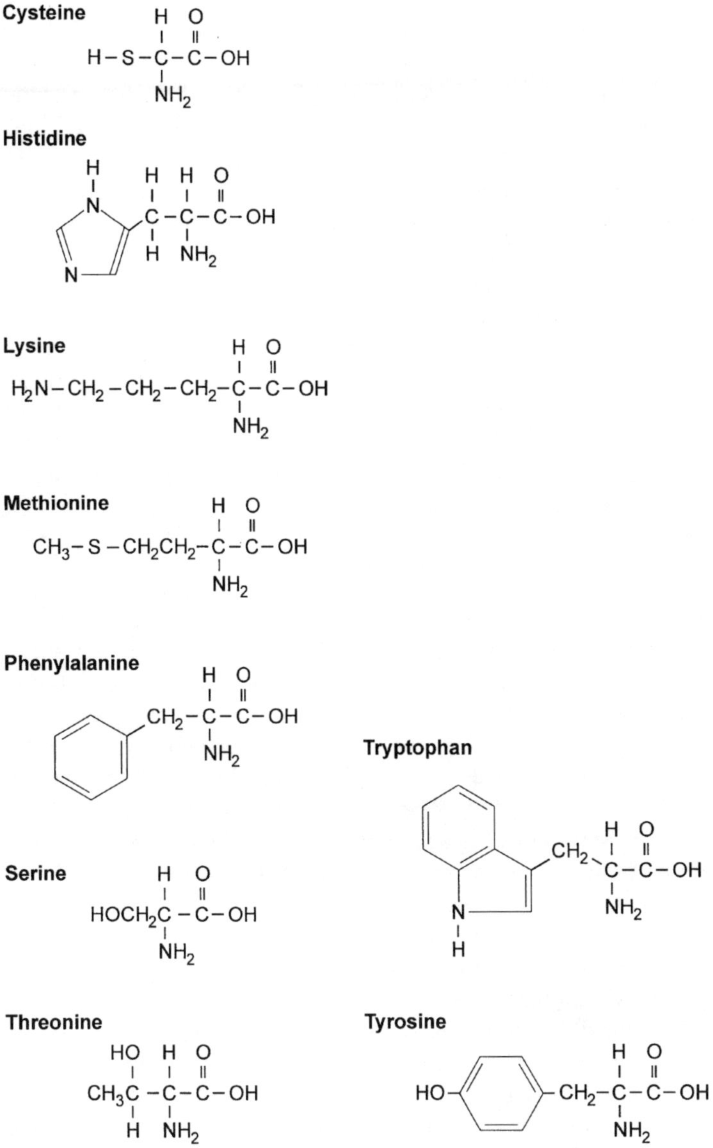

Figure 3.1. Structures of amino acids that are susceptible to oxidative damage

generated by the oxidation of serine or threonine, or with quinones formed
by the hydroxylation and oxidation of the aromatic rings of phenylalanine,
tyrosine, tryptophan or histidine. Active-site lysines of gastric peroxidase
have been reported to be oxidised by hydroxyl radicals (Das *et al.* 1998).

Tyrosine and phenylalanine

Polymerisation of tyrosines to form the pigment melanin is a well-known and widely observed phenomenon. The reaction is catalysed by the enzyme tyrosinase and requires the presence of oxygen. The first step of the reaction is the addition of another hydroxyl group to form hydroxytyrosine, followed by the oxidation of the phenolic hydroxyls to quinones. The quinones can then form covalent bonds with an adjacent phenol or amine group. In the chemically catalysed reaction, the phenyl ring is hydroxylated and then spontaneously oxidises to the quinone form. The quinone can then form a covalent adduct with an electron donor such as a hydroxyl, thiol or amino group. The effect of dimerisation of tyrosines on the activity of an enzyme is quite variable and depends on the type of enzyme. Yeast invertase was found to be quite sensitive to both enzymatic and chemically catalysed tyrosine dimerisation, while chymotropsin, trypsin and pepsin showed no loss of activity (Sizer 1946, 1947, 1948; Wiseman and Woodward, 1974). Hydroxyl radicals generated by reaction of Ni(II)-histidine have been shown to catalyse the formation of bityrosine in model peptides (Torreilles and Guerin 1990).

Tryptophan and histidine

The indole ring of tyrptophan can be hydroxylated by hydroxyl ion, leading to ring cleavage and formation of N-formylkynurenine and kynurenine. This destruction of the indole ring can be followed by the loss of fluorescence of tryptophan during the oxidation. Tryptophan oxidation has been detected as a result of oxidative damage in bovine α-crystallin (Finley *et al.* 1997) and as a result of Ni(II)-histidine-catalysed formation of hydroxyl radicals in model peptides (Torreilles and Guerin 1990). Hydroxyl radicals generated from copper salts and hydrogen-peroxide caused loss of tryptophan fluorescence of albumin, γ-globulin, and papain (Gutteridge and Wilkins 1983). Oxidation and carbonyl condensation products were identified in tryptophan lots implicated in a novel medical condition, eosinophilia–myalgia syndrome (Simat *et al.* 1996).

In the cellulose-binding domains (CBDs) of cellulases and xylanases, tryptophan residues form important contacts with crystalline cellulose (Poole *et al.* 1993; Linder and Teeri 1997). Specific chemical oxidation of two tryptophans in the CBD in the glucanase–xylanase Cex from *Cellulomonas fimi* was reported to reduce the binding of the CBD to crystalline cellulose by 90% (Bray *et al.* 1996).

Histidine can be hydroxylated by ROS by mechanisms similar to those for the other aromatic amino acids. Since histidine can bind metal ions, and indeed is often a coordinating ligand in metalloenzymes, it is susceptible to autocatalysed oxidative inaction (Gutteridge and Wilkins 1983; Torreilles and Guerin 1990; Zhao *et al.* 1996, 1997). Of the metal ions that are known to bind to histidine, nickel, iron, copper and ruthenium have been shown to catalyse formation of hydroxyl radicals by the Fenton reaction (Aust *et al.* 1985; Meinecke *et al.* 1996).

Methionine, serine and threonine

Methionine can be oxidised by hydrogen peroxide to form methionine sulphoxide (Kido and Kassel 1975). This reaction is reversible by treatment with thiol reagents such as thioglycolic acid, cysteine, mercaptoethanol or dithiothreitol for 18 hours (Dedman *et al.* 1957, 1961; Houghten and Li 1983). The enxyme methionine sulphoxide reductase catalyses the reduction of methionine sulphoxide to methionine in the presence of reduced thioredoxin or dithiothreitol (Abrams *et al.* 1981).

Reactive oxygen groups are known to oxidise the hydroxyl groups of serine and threonine to aldehydes and ketones. These carbonyls can then react with amino groups or other protein moieties, or with the hydroxyl groups of glycans, to form crosslinks (Traverso *et al.* 1997; Yan *et al.* 1998).

Inactivation of metalloenzymes

Certain groups of metalloenzymes are intrinsically sensitive to inactivation by oxidation. These include members of the iron–sulphur proteins, in which the iron is coordinated to four or more cysteinyl thiols (Beinert *et al.* 1997). Reversible hydrogenases that are able to produce hyrodgen as well as oxidise it have been found to be highly oxygen sensitive (Lindmark and Muller 1973; Adams and Hall 1977; Erbes *et al.* 1979; Adams 1990). An algal hydrogenase from *Chlamydomonas reinhardtii* was found to have a second-order rate constant for irreversible inhibition by oxygen of $1.4 \times 10^5 M/$ minute (Erbes *et al.* 1979). Hydrogenases that solely catalyse hydrogen uptake are not as oxygen sensitive as the reversible hydrogenases (Hahn and Kück 1994; Adams and Hall 1977; Schneider *et al.* 1979). The hydrogenase from the non-sulphur purple bacterium *Rhodospirillum rubrum*, an iron–sulphur protein, is relatively stable under aerobic conditions at low temperatures such as $-20\,°C$ (Adams and Hall 1977), while the hydrogenase from *Alcaligenes eutrophus* is reported not to be inhibited by oxygen (Schneider *et al.* 1979).

Enzymes containing heme groups, such as lignin peroxidase, offer special problems. They require dissolved oxygen species for activity, often provided by adding hydrogen peroxide to the reaction mixtures. However, too high a concentration of oxygen leads to destruction of their catalytic activity (Tien and Kirk 1984; Tuisel *et al.* 1990). Oxidative inactivation of lignin peroxidase has been linked to hydroxylation of an active-site tryptophan residue (Blodig *et al.* 1998). Adding free tryptophan to the media used to grow white rot fungi stabilised lignin peroxidase against hydrogen peroxide-catalysed oxidation; tryptophan was found to be a better substrate for the peroxidate than veratryl alcohol, its natural substrate (Collins *et al.* 1997). Extrinsic factors such as other redox proteins can stabilise oxygen-sensitive metalloenzymes such as nitrogenase (Robson 1979).

Antioxidants and their application (see Chapters 1, 6 and 7)

Inert or reducing atmosphere

Perhaps the most obvious, if not always the most convenient, method of avoiding oxidative damage is by removing and excluding oxygen from the microenvironment and buffers being used for the preparation and storage of the proteins. Dissolved oxygen can be removed to some extent by degassing of solutions under vacuum, but the preferred method is the replacement of dissolved oxygen by sparging the solution with an inert gas such as nitrogen, argon or helium. A mixture of nitrogen, argon, or helium and hydrogen can be used to provide a reducing atmosphere for the culture media or reaction. Undiluted hydrogen gas was used to maintain a reducing atmosphere for the purification of hydrogenase from *Clostridium pasteurianum* (Nakos and Mortenson 1971). Since hydrogen is considered flammable at concentrations higher than 3% by volume, appropriate safety procedures must be implemented when hydrogen gas is used. Oxygen-sensitive reactions can be conducted inside glove boxes or bags or in sealed vessels with the headspace filled with nitrogen or nitrogen/carbon dioxide (Burdette and Zeikus 1994; Bryant and Adams 1989; Aono *et al.* 1989; Adams 1990; Adams and Zhou 1995; Zhang *et al.* 1996).

Hydrogen sulphide is used in the syntheses of mercaptoethanol and other thiol compounds. It is also a product of microbial degradation of sulphur-containing compounds. Since it is a gas that is both toxic (Part 261 of Title 40, Code of Federal Regulations 261 USA; OSHA Short-Term Exposure Limit, 15 ppm) and explosive (flammable limit 4% by volume), the proper detection and disposal systems must be provided if hydrogen sulphide gas is being used or generated (Beauchamp *et al.* 1984; Baross 1995). Obviously, its use in routine protein purification requires glove boxes and toxic gas detectors. Sodium sulphide (Baross, 1995) and ferrous sulphide (Brock and Od'ea 1977) are disulphides that can be used to generate a reducing atmosphere in anaerobic culture media by *in situ* production of hydrogen sulphide.

To determine whether oxygen has been removed from solutions, resazurin (7-hydroxy-3H-phenoxazine-3-one 10-oxide) can be added at a concentration of 0.001 g/l (0.1%) to media for anaerobes. This redox indicator is pink in oxygenated solutions, but turns colourless as oxygen is removed (Baross 1995). When carrying out analyses of anaerobic cultures containing resazurin, the redox properties of this reagent should be considered. Since resazurin itself is a redox agent, it can be enzymatically reduced by lactate dehydrogenase (Jacobsen et al. 1987) and NADH oxidoreductase (Barnes and Spenny 1980). Resazurin has been reported to form free radicals capable of reaction with NADH, reduced glutathione and hydroxytyrosine under aerobic conditions (Prutz et al. 1996).

Another method to protect an oxidation-sensitive enzyme utilised a 'salting-out' effect to reduce the local oxygen concentration. A clostridial hydrogenase was stabilised by attachment to a polycationic surface (Klibanov et al. 1978).

Cysteine and glutathione

The reducing compounds that most closely approximate the thiol groups of proteins are the amino acid cysteine and the tripeptide glutathione (L-γ-glutamyl-L-cysteinyl L-glycine). Glutathione is the major antioxidant found in cells. It is present at concentrations ranging from 0.1 to 10 mM in animal cells. Reduced glutathione concentrations are maintained in the cell by the action of glutathione disulphide reductase. This enzyme couples reduction of oxidised glutathione to NADPH oxidation. Glutathione reacts directly with free radicals, but also serves to detoxify many drugs and other reactive compounds as a substrate of glutathione s-transferase (Meister 1995). Cysteine and glutathione have the advantages of low toxicity and high melting point, causing less odour than other mercaptans. When oxidised to the dimeric forms, cysteine and oxidised glutathione, the solubility of these compounds greatly decreases. The application of glutathione or cysteine to the activation of the sulphydryl proteases papain and cathepsin is well known (Bersin and Logemann 1933; Grassmann et al. 1931). Glutathione and cysteine can be used to assist the correct refolding of reduced, denatured proteins (Raman et al. 1996).

Thiol reagents

The two reducing agents most familiar to biochemists are 2-mercaptoethanol and dithiothreitol (Figure 3.2). These compounds can be used to protect proteins from oxidation, but their reducing action can also result in protein inactivation. Concentrations of 0.2% or 14 mM 2-mercaptoethanol or 2 mM dithiothreitol are commonly recommended as working concentrations in

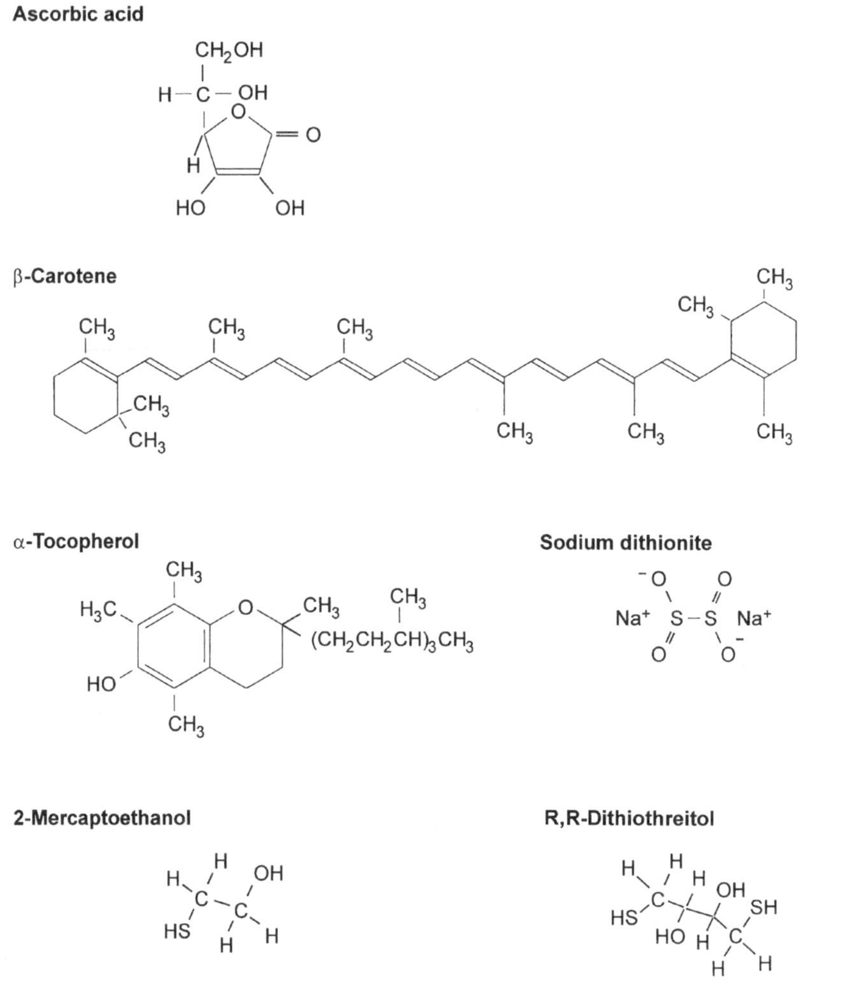

Ascorbic acid

β-Carotene

α-Tocopherol

Sodium dithionite

2-Mercaptoethanol

R,R-Dithiothreitol

Figure 3.2. Structures of common antioxidants

buffers (Kaufman *et al.* 1995). At higher concentrations, they are used to reduce disulphide bonds to unfold and denature proteins. Proper refolding to regenerate an active enzyme is dependent on protein concentration, thiol concentration, and other factors (Anfinsen *et al.* 1961; Anfinsen and Haber 1961; White 1961). Dithiothreitol can cause displacement of iron from phenylalanine hydroxylase (Martinez *et al.* 1992). To completely denature proteins for polyacrylamide gel electrophoresis, 5% 2-mercaptoethanol and 2.5% sodium dodecyl sulphate are often used (Laemmli 1970).

The synthesis of 2-mercaptoethanol (also known as β-mercaptoethanol, thioglycol, monothioethylene glycol, 2-hydroxyethanethiol) from ethylene chlorohydrin and sodium sulphide was reported in 1862 by Carius. A new, more efficient synthesis used ethylene oxide and hydrogen sulphide (Woodward 1948). This compound is well known to have an offensive odour. The toxicological properties of undiluted 2-mercaptoethanol require that solutions should be prepared in fume hood and that safety glasses and gloves be worn during use.

Dithiothreitol (DDT; R,R-1,4-dithiothreitol; threo-2,3-dihydroxy-1,4-thiolbutane) is often called Cleland's reagent. The synthesis of dithiothreitol and dithioerythritol had been reported some 20 years earlier (Evans et al. 1949), but it was W. W. Cleland who first realised that such a dithiol compound would offer several advantages for maintaining protein thiol groups in the reduced form. In addition to water solubility and reduced stench, he thought that less of the thiol compound would be required to maintain protein thiols in the reduced state. Basically, the capacity of dithiothreitol to form an intramolecular disulphide bond causes its oxidation to be thermodynamically more favourable than in the case of a monothiol that could form only an intermolecular disulphide. For a monothiol, the equilibrium for reduction of a protein thiol would be near 1. In the course of his investigations, Cleland found the equilibrium constant for cyclisation of dithiothreitol to be about 10 and the equilibrium constant for reduction of cystine by dithiothreitol to be 1.3×10^4. The properties of dithiothreitol are basically the same as those of its isomer dithiothreitol (Cleland 1964).

Sodium dithionite

Sodium dithionite (Fig. 3.2) is used industrially as a reducing agent in indigo staining of denim and other processes. It can be added to buffers at 2 mM concentration to remove traces of oxygen for purification or assays of extremely oxygen-sensitive proteins such as hydrogenase from P. furiosus (Peck and Guest 1955; Adams and Hall, 1977; Adams and Zhou, 1995).

Thiols can indirectly cause oxidative damage. At low concentrations, dithiothreitol and other thiol-containing compounds can react with oxygen and transition metals such as iron or copper to produce hydrogen peroxide. The hydrogen peroxide then reacts with metal ions to form hydroxyl radicals. At higher concentrations, enough of the dithiothreitol itself is present to act as a radical sink to reduce the hydroxyl radicals. In the case of V79 Chinese hamster lung fibroblasts, 0.7 mM dithiothreitol was toxic, but 2.0 mM prevented cell death (Held et al. 1996).

Ascorbic acid

Ascorbic acid (vitamin C) is a well-known biological antioxidant (Fig. 3.2), which plants and many animals are able to synthesise. It is an essential nutrient for humans and other primates, who lost the ability to carry out its synthesis an estimated 25 million years ago. Although the nutritional effects of plant products such as fresh fruits and vegetables in prevention of scurvy had been observed for some 300 years, it was not until 1932 that Albert Szent-Györgyi isolated and crystallised ascorbic acid from red peppers. Using this purified sample, the structure was determined by W. M. Haworth. Soon pure crystalline ascorbic acid became commercially available and is still being investigated for its role in the treatment of health problems ranging from rhinovirus infections to rheumatism (Pauling 1970). Increased levels of ascorbic acid are an indicator of oxidative stress in biological tissues (Buettner and Jurkiewicz 1996). Elevated levels of ascorbic acid were detected in white muscle of dystrophic chickens (Perkins *et al.* 1980). Ascorbic acid has been reported to repair oxidative damage *in vitro* by reducing tryptophan and tyrosine radicals in free amino acids and in lysozyme (Hoey and Butler 1984).

Ascorbic acid contains a lactone ring that cycles between the reduced enol form (ascorbic acid) and the oxidised keto form (dehydroascorbic acid). The redox potential is influenced by pH (Table 3.2).

Since hydrolysis of the lactone ring, which renders the compound more susceptible to degradation, is favoured at pH values $\leqslant 7$, ascorbic acid is most stable at acidic pH. Analysis of the products of degradation of ascorbic acid by hydrogen peroxide identified diketogluonic acid and tetrahydroxy-diketohexanoic acid (Deutsch 1998).

Although it is widely used as a reducing agent and antioxidant, ascorbic acid can also act as a pro-oxidant. It is implicated as a participant in oxidative damage due to hydroxyl radical production. Since ascorbic acid can reduce transition-metal ions such as iron, copper, nickel and ruthenium, under aerobic conditions it can catalyse Fenton reactions if such metals are present:

Table 3.2. Effect of pH on redox potential of ascorbic acid/dehydroascorbic acid couple[a]

Reaction pH	E^0
4	0.166
7	0.058
8.7	−0.012

[a]Loach 1968.

$$\text{Ascorbic acid} + Fe^{3+} \rightarrow \text{dehydroascorbic acid} + Fe^{2+}$$

$$Fe^{2+} + H_2O_2 \rightarrow Fe^{3+} + 2\,^{\bullet}OH$$

Copper is 80 times more efficient than iron as a catalyst for ascorbate oxidation. Traces of iron at $1\,\mu M$ or copper at $0.1\,\mu M$ are sufficient to produce detectable levels of hydroxyl radicals in reaction with ascorbic acid. This is known as the crossover effect: at high concentration, ascorbate acts as an antioxidant; at low concentration, as a pro-oxidant, assuming a low-metal ion concentration (Buettner and Jurkiewicz 1996). Hydroxyl radicals formed by the Fenton reaction hydroxylate aromatic rings of protein amino acids including histidine and tryptophan (Atalay *et al.* 1998; Delgado and Slobodian 1972), and cause formation of tyrosine dimers (Torreilles and Guerin 1990). Oxidative deamination and decarboxylation of amino acids have also been reported (Stadtman and Berlett 1991). Ascorbate has been demonstrated to reversibly inhibit catalase (Davison *et al.* 1986).

Tocopherols

Another well-known family of nutritional antioxidants are the tocopherols (vitamin E). Tocopherols (Fig. 3.2) are antioxidants for vitamin A and its precursor β-carotene (Stern *et al.* 1947). They protect membranes from oxidative damage, forming a complex with fatty acids by hydrogen bonds between the hydroxyl groups of tocopherol and the carboxyl groups of linoleic acid (Erin *et al.* 1985). There is great interest in vitamin E as a therapeutic treatment, particularly for neurological degeneration. Such treatment is reported to delay the onset of familial amyotrophic lateral sclerosis, a condition caused by increased hydroxyl ion production due to mutations in superoxide dismutase (Price *et al.* 1998). However, treatment of patients with Parkinson's disease showed no lasting effect (Shoulson 1998). Genetic engineering of plants to increase the fraction of the most important isomer, α-tocopherol, is being carried out (Shintani and DellaPenna 1998). Application of tocopherols as antioxidants is constrained by their bulkiness and hydrophobicity. In amino acid repair experiments, α-tocopherol quickly reduced free tryptophan radicals, but was able only to slowly repair oxidised tryptophan residues of lysozyme (Hoey and Butler 1984). In culture media α-tocopherol protected *E. coli* cells from superoxide radical damage, the vitamin being incorporated into the cell membranes (Fuentes and Amabile-Cuevas 1998). Synergistic effects are observed when combinations of tocopherol and ascorbic acid are used as antioxidants (Fuentes and Amabile-Cuevas 1998). Tocopherols provide protection for lipid-containing structures such as cell membranes, while water-soluble ascorbic acid reduces oxidants in the aqueous environment. Additionally, ascorbic acid is able to repair oxidative damage to tocopherols by reducing the tocopherol radical (Buettner and Jurkiewicz 1996).

β-Carotene

The carotenoids, in particular β-carotene (Fig. 3.2), are being intensively investigated as treatments for diseases such as arteriosclerosis and cancer (Jacob and Burri 1996). Like other carotenoids, β-carotene plays an important role in photosynthetic plants (Young *et al.* 1997). It acts as a scavenger of peroxy and phenoxyl radicals (Canfield *et al.* 1992; Ozhagina and Kasaikina 1995; Tyurin *et al.* 1997), is lipophilic, and is not very soluble in water. Physiological concentrations are reported to be about 0.5 nmol/l; concentrations of 3 to 36 nmol/l in cell culture media were achieved by using an ethanol solution (Gross *et al.* 1997). It carries out its antioxidant function when incorporated into a membrane or lipid layer (Tsuchihashi *et al.* 1995). Basically, it prevents oxidative damage by preventing lipid oxidation (Kennedy and Liebler 1992). Consequently, it is likely to only protect proteins that are incorporated into a lipid layer or that are highly hydrophobic, such as low-density lipoproteins (Jialal *et al.* 1991; Carpenter *et al.* 1997).

Enzymatic methods

A number of enzymes carry out reduction of oxygen or other ROS either as their primary function or coupled to the oxidation of a substrate. Such enzymes can be used to establish and maintain an anaerobic micro-environment.

The enzyme glucose oxidase couples reduction of oxygen to oxidation of glucose to produce hydrogen peroxide and gluconic acid as products:

$$O_2 + glucose + H_2O \rightarrow H_2O_2 + gluconic\ acid$$

Glucose oxidase and glucose were used as an oxygen trap to remove oxygen from a hydrogen-producing system consisting of spinach ferro-doxin and spinach chloroplasts in a nitrogen atmosphere (Rao *et al.* 1976). The products of the glucose oxidase reaction affect the reaction micro-environment by a drop in pH due to gluconic acid production and by possible inhibition due to hydrogen peroxide. The glucose oxidase–glucose system inactivated glutamine synthetase from *E. coli* (Stadtman and Wittenberger 1985). Glucose oxidase can be immobilised on a number of different supports (Markey *et al.* 1975).

Two cellular enzymes that act to catalyse the decomposition of ROS are superoxide dismutase and catalase. A survey of micro-organisms for presence of these two enzymes found that the amounts of superoxide dismutase and catalase activity present were directly related to the oxygen tolerance of the micro-organism. Strict anaerobes did not possess either enzyme, while aerotolerant anaerobes demonstrated only superoxide dismutase activity and aerobes had high levels of both enzymes (McCord *et al.* 1971). Superoxide dismutase catalyses the breakdown of superoxide anion to molecular oxygen (McCord and Fridovich 1968; Steinman *et al.* 1974):

$$2O_2^- + 2H^+ \rightarrow H_2O_2 + O_2$$

Catalase catalyses the decomposition of hydrogen peroxide to water and molecular oxygen (Zidoni and Kremer 1974):

$$H_2O_2 \rightarrow H_2O + \tfrac{1}{2}O_2$$

Both of these enzymes can themselves be inactivated by ROS at sufficiently high concentration, and, indeed, the products of the reactions that they catalyse are also reactive. Human superoxide dismutase mutants that evidenced higher activity than the wild type were found to become hydroxylated, implying that a lower reaction rate preserves enzyme activity (Wiedau-Pazos *et al.* 1996). Toxicity can be reduced if superoxide dismutase and catalase are both used in the system. A combination of catalase and superoxide dismutase was reported to function as a long-lasting antioxidant system in cell culture (Mahan and Insel 1984). A superoxide dismutase–catalase conjugate was found to be a less toxic antioxidant than free superoxide dismutase for prevention of reperfusion injury in tissues (Mao *et al.* 1993).

Oxyrase™ is an oxygen-reducing membrane suspension isolated from a selected *E. coli* strain by centrifugation and sterilised by filtration through a 0.22 μm filter. The product contains several membrane-bound enzymes and ubiquinone. By addition of suitable hydrogen donors such as lactate and succinate at 5 to 50 mM concentrations, the preparation can be used to generate anaerobic conditions in culture media or biochemical solutions by the reaction of the membrane-bound dehydrogenases with their substrates to reduce oxygen (Adler and Spady 1977).

The major disadvantage of the use of redox enzymes to scavenge oxygen radicals appears to be the possibility of unforeseen effects on the systems being protected that can complicate interpretation of experimental results. Possible effects of the reaction products from the oxygen-removing enzyme should be considered when designing an experiment.

Table 3.3. Effect of pretreatment with ascorbic acid on specific activity of native CBH I and pentaammineruthenium–CBH I on filter paper[a]

Enzyme	Control	Ascorbic acid treated
Native CBH I	0.226	0.284
Pentaammineruthenium–CBH I	0.218	0.268

[a]Specific activity is given in μmol reducing sugar (mg protein/minute) for incubation with 1% filter paper in 50 mM sodium acetate, pH 4.3 at 44 8C.

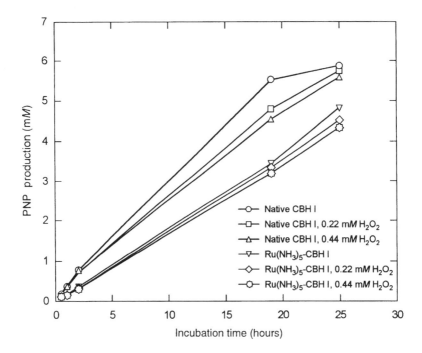

Figure 3.3. Effect of hydrogen peroxide on hydrolysis of 10 mM p-nitrophenyl β-D-cellobioside (PNPC) to p-nitrophenol (PNP) and cellobiose at pH 4.3 and 55 °C by native CBH I and pentaammineruthenium–CBH I

Two case studies of effects of oxidants

Stability of a fungal cellulase to oxidation

At this point in the discussion, one may well ask how serious is the problem of protein oxidation in a given circumstance. The answer depends on the protein of interest and its stability. As an example, to create a bifunctional

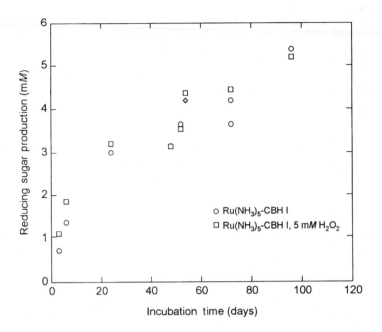

Figure 3.4. Effect of 5 mM hydrogen peroxide on hydrolysis of 1% newsprint at pH 4.3 and 42 °C by pentaammineruthenium–CBH I

enzyme capable of oxidising lignin as well as hydrolysing cellulose, pentaammineruthenium complexes were attached to cellobiohydrolase I (CBH I), the major cellulase produced by the fungus *Trichoderma reesei* (Evans *et al.* 1993, 1994, 1995). The pentaammineruthenium–CBH I construct was able to carry out hydrolysis of cellulosic substrates as well as oxidise veratyl alcohol in the presence of hydrogen peroxide. Pentaammineruthenium–CBH I was shown to oxidise veratryl alcohol at hydrogen peroxide concentrations of 1 to 4 mM (Evans *et al.* 1994). Although ruthenium has been reported to form hydroxyl radicals in a Fenton's reaction with ascorbic acid and hydrogen peroxide (Meinicke *et al.* 1996), pentaammineruthenium–CBH I that had been pretreated with 2 mM ascorbic acid suffered no loss of activity on crystalline cellulose (Table 3.3).

The effect of hydrogen peroxide on hydrolysis of the synthetic substrate *p*-nitrophenyl β-D-cellobioside was examined for native and pentaammine-ruthenium-modified CBHI. No decrease in hydrolysis rate was observed for 0.22 or 0.44 mM hydrogen peroxide (Figure 3.3). Hydrolysis of cellulosic substrates was also examined. The hydrolysis of newsprint by pentaammineruthenium–CBH I did not appear to be affected by 5 mM

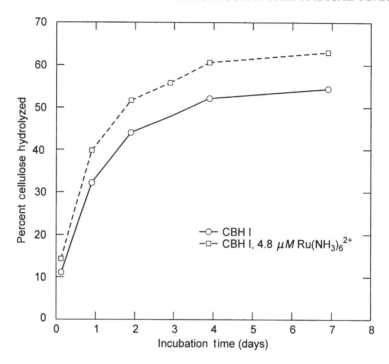

Figure 3.5. Effect of hexaammineruthenium(II) on hydrolysis of 1% microcrystalline cellulose (Avicel) by CBH I and *Aspergillus niger* β-glucosidase at pH 4.3 and 42 °C

hydrogen peroxide (Figure 3.4). In another experiment, native CBH I was incubated for seven days at 42 °C and pH 5 with 1% crystalline cellulose and Novozym 188 β-glucosidase in the presence of 4.8 μM hexaammine-ruthenium chloride that had been reduced with a twofold excess of ascorbic acid. No decrease in hydrolysis rate in the presence of hexaammine-ruthenium(II) was observed (Figure 3.5).

Table 3.4. Specific activities in μmol (μmol catalyst)/minute of free MP8 and MP8-modified CBH I for oxidation of aromatic substrates[a]

Substrate	MP8	MP8-modified CBH I
Hydroquinone	48.24	21.58
Veratryl alcohol	0.792	0.454
Tyrosine	28.3	8.69

[a]Reactions were carried out at 23 °C and pH 7.0 in 100 mM sodium phosphate buffer with 1.0 mM hydrogen peroxide and 0.5 μM MP8 or 1.0 μM MP8-modified CBH I (0.4 MP8 attached per CBH I).

Table 3.5. Hydrolytic activity of native and MP8-modifed CBH I[a]

Substrate	Native CBH I	MP8-modified CBH I
10 nM PNPC	0.0415	0.0481
0.5% Barley β-glucan	0.925	0.977
1% Avicel (reducing sugar)	0.266	0.258
1% Avicel (glucose)	0.0120	0.0104

[a]Specific activities are given in μmol (mg protein)/minute at 30 minute incubation for p-nitrophenyl β-D-cellobioside (PNPC) and glucan and at 3 hours for Avicel. Reactions were carried out at 45 °C and pH 4.3 in 50 mM sodium acetate buffer, with substrate concentrations as indicated and 0.0040 mg/ml enzyme.

Pentaammineruthenium is a less effective oxidiser than other metal complexes such as hemes (Table 3.1). To improve the redox capability of the metallocellulase construct, a heme peptide was used instead of penta-ammineruthenium. In previously unpublished experiments carried out in 1994 and 1995 (Evans, Margalit, and Woodward, unpublished), micro-peroxidase-8 (MP8), a heme peptide eight amino acids long obtained from digestion of horse heart cytochrome c by pepsin and trypsin (Harbury and Loach 1960), was attached to CBH I using water-soluble carbodiimide. To protect the active site of the CBH I from modification by the carbodiimide, 100 mM cellobiose was included during the modification. MP8 carries out hydrogen peroxide-dependent oxidation of guaiacol (Baldwin et al. 1985) and o-dianisidine (Kraehenbuhl et al. 1974) as well as hydroxylation of aromatic rings and oxidation of phenolic hydroxyls to quinones (Osman et al. 1996). The bifunctional MP8–CBH I construct was found to have both the redox activity pertaining to MP8 (Table 3.4) as well as the cellulase activity of CBH I (Table 3.5).

It is noteworthy that no loss in activity on crystalline cellulose was observed, even though hydrolysation of important tryptophan residues of the cellulose binding domain could have been carried out by the attached MP8. Since *T. reesei* and other fungal cellulases function in aerobic environments, it is not surprising to find that these enzymes are resistant to inactivation by oxidation. Not only are these aerobic micro-organisms, wood-rot fungi are known to produce enzymes such as cellobiose oxidase that generate the hydrogen peroxide needed by the lignin peroxidase and manganese peroxidase that depolymerise the lignin of the lignocellulosic substrate (Kremer and Wood 1992a,b). Glycosylation has been reported to protect proteins from oxidative damage (West 1986; Rudd et al. 1994; Uchida et al. 1997). CBH I is 5–10% glycosylated in the linker region connecting the catalytic domain to the cellulose-binding domain (Tomme et al. 1988; Esterbauer et al. 1991). It is probable that glycosylation and other structural characteristics of CBH I and other cellulases render them resistant to incapacitating oxidative damage.

Use of hydrogenase for production of hydrogen

Even extremely oxidation-sensitive enzymes can be used under ordinary laboratory conditions. Although the purification of P. furiosus hydrogenase must be carried out under stringent exclusion of oxygen, the enzyme retains sufficient activity under aerobic conditions for assays to be carried out for short periods of time, approximately five minutes (Hershlag et al. 1998). For longer incubation periods ranging from a few hours to several days, an inert atmosphere of helium gas must be employed in the reaction vessel (Woodward et al., 1996).

References

Abrams, W.R., Weinbaum, G., Weissbach, L., Weissbach, H. and Brot, N. (1981) Enzymatic reduction of oxidised α-1-proteinase inhibitor restores biological activity. Proc. Natl. Acad. Sci. USA, **78**, 7483–7486.

Adams, M.W.W. (1990) The structure and mechanism of iron-hydrogenases. Biochim. Biophys. Acta, **1020**, 115–145.

Adams, M.W.W. and Hall, D.O. (1977) Isolation of the membrane-bound hydrogenase from Rhodosprillum rubrum. Biochem. Biophys. Res. Commun., **77**, 730–737.

Adams, M.W.W. and Zhou, Z.H. (1995) Protocol 9. Purification of hydrogenase and ferrodoxin from a hyperthermophile. In: Archaea: Thermophiles. A Laboratory Manual, Cold Spring Harbor Laboratory Press, Danvers, MA, F.T. Robb and A.R. Place, eds. pp. 67–71.

Adler, H. and Spady, G. (1977) The use of microbial membranes to achieve anaerobiosis. J. Rapid Methods Automation Microbiol. **5**, 1–12.

Anfinsen, C.B. and Haber, E. (1961) Studies on the reduction and reformation of protein disulfide bonds. J. Biol. Chem., **236**, 1361–1363.

Anfinsen, C.B., Haber, E., Sela, M. and White, F.H. (1961) The kinetics of formation of native ribonuclease during oxidation of the reduced polypeptide chain. Proc. Natl. Acad. Sci. USA, **47**, 1309–1314.

Aono, S., Bryant, F.O. and Adams, M.W.W. (1989) A novel and remarkably thermostable ferredoxin from the hyperthermophilic archaebacterium Pyrococcus furiosus. J. Bacteriol., **171**, 3433–3439.

Atalay, A., Ogus, A., Bateman, O. and Slingsby, C. (1998) Vitamin C-induced oxidation of eye lens gamma crystallins. Biochemie **80**, 283–288.

Aust, S.D., Morehouse, L.A. and Thomas, C.E. (1985) Role of metals in oxygen radical reactions. J. Free Radical. Biol. Med., **1**, 3–5.

Baldwin, D.A., Marques, H.M. and Pratt, J.M. (1985) Mechanism of activation of H_2O_2 by peroxidases: Kinetic studies on a model system. FEBS Lett., **183**, 309–312.

Balls, A.K., Lineweaver, H. and Thompson, R.R. (1937). Crystalline papain. Science, **86**, 379.

Barnes, S. and Spenny, J.G. (1980) Stoichiometry of the NADH-oxidoreductase reaction for dehydrogenase determinations. Clin. Chim. Acta, **107**, 149–154.

Baross, J.A. (1995) Protocol 2. Isolation, growth, and maintenance of hyperthermophiles. In: *Archaea: Thermophiles. A Laboratory Manual*, F.T. Robb and A.R. Place, eds, Cold Spring Harbor Laboratory Press, Danvers, MA, pp. 19–23.

Beauchamp, R.O., Jr, Bus, J.S., Popp, J.A., Boreiko, C.J. and Andjelkovich, D.A. (1984) A critical review of the literature on hydrogen sulfide toxicity. *CRC Crit. Rev. Toxicol.*, **13**, 25–97.

Beinert, H., Holm, R.H. and Münck, E. (1997) Iron–sulfur clusters: Nature's modular, multipurpose structures. *Science*, **277**, 653–659.

Bersin, T. and Logemann, W. (1933) Influence of oxidizing agents and reducing agents on the activity of papain. *Ztschr. Physiol. Chemie*, **220**. 209–216.

Blodig, W., Doyle, W.A., Smith, A.T., Winterhalter, K., Choinowski, T. and Piontek, K. (1998) Autocatalytic formation of a hydroxy group at C beta of Trp171 in lignin peroxidase. *Biochemistry*, **37**, 8832–8838.

Boyer, P.D. (1960) *The Enzymes*, Vol. 4, 2nd edn. Academic Press, New York, pp. 233–241.

Bray, M.R., Johnson, P.E., Gilkes, N.R., McIntosh, L.P., Kilburn, D.G. and Warren, R.A. (1996) Probing the role of tryptophan residues in a cellulose-binding domain by chemical modification. *Protein Sci.*, **5**, 2311–2318.

Brock, T.D. and Od'ea, K. (1977) Amorphous ferrous sulfide as a reducing agent for culture of anaerobes. *Appl. Environ, Microbiol.*, **33**, 254–256.

Bryant, F.O. and Adams, M.W.W. (1989) Characterization of hydrogenase from the hyperthermophilic archaeobacterium *Pyrococcus furiosus*. *J. Biol. Chem.*, **2643**, 5070–5079.

Buettner, G.R. and Jurkiewicz, B.A. (1996) Catalytic metals, ascorbic acid and free radicals: Combinations to avoid. *Radiat. Res.*, **145**, 532–541.

Burdette, D.S., Secundo, F., Phillips, R.S., Dong, J., Scott, R.A. and Zeikus, J.G. (1997) Biophysical and mutagenic analysis of *Thermoanaerobacter ethanolicus* secondary-alcohol dehydrogenase activity and specificity. *Biochem. J.*, **326**, 717–724.

Burdette, D. S. and Zeikus, J.G. (1994) Purification of acetaldehyde dehydrogenase and alcohol dehydrogenases from *Thermoanaerobacter ethanolicus* 39E and characterization of the secondary alcohol dehdyrogenase (20 Adh) as a bifunctional alcohol dehydrogenase-Acetyl CoA reductive thioesterase. *Biochem. J.*, **302**, 163–170.

Canfield, L.M., Forage, J.W. and Valenzuela, J.G. (1992) Carotenoids as cellular antioxidants. *Proc. Soc. Exp. Biol. Med.*, **200**, 260–265.

Carpenter, K.L., van der Veen, C., Hird, R., Dennis, I.F., Ding, T. and Mitchinson, M.J. (1997) The carotenoids β-carotene, canthaxanthin, and zeaxanthin inhibit macrophage-mediated LDL oxidation. *FEBS Lett.*, **401**, 262–266.

Cleland, W.W. (1964) Dithiothreitol, a new protective reagent for SH groups. *Biochemistry*, **3**, 480–482.

Collins, P.J., Field, J.A., Teunissen, P. and Dobson, A.D. (199) Stabilization of lignin peroxidase in white rot fungi by tryptophan. *Appl. Environ. Microbiol.*, **63**, 2543–2548.

Das, D., Bandyopadhyay, D. and Banerjee, R.K. (1998) Oxidative inactivation of gastric peroxidase by site-specific generation of hydroxyl radical and its role in stress-induced gastric ulceration. *Free Radical Biol. Med.*, **24**, 460–469.

Davison, A.J., Kettle, A.J., Fatur, D.J. (1986) Mechanism of the inhibition of catalase by ascorbate. Roles of active oxygen species, copper, and semidehyroascorbate. *J. Biol. Chem.*, **261**, 1193–1200.

Dedman, M.L., Farmer, T.H. and Morris, C.J.O.R. (1957) Studies on pituitary adrenocorticotropin: 2. Oxidation–reduction properties of the hormone. *Biochem. J.*, **78**, 348–352.

Dedman, M.L., Farmer, T.H. and Morris, C.J.O.R. (1961) Studies on pituitary adrenocorticotropin: 3. Identification of the oxidation–reduction centre. *Biochem. J.*, **78**, 348–352.

Dekker, K., Yamagata, H., Sakaguchi, K. and Udaka, S. (1991) Xylose (glucose) isomerase gene from the thermophile *Clostridium thermohydrosulfuricum* 39E and characterization of the enzyme purified from *Escherichia coli*. *Agric. Biol. Chem.*, **55**, 221–227.

Delgado, C.J. and Slobodian, E. (1972) The inactivation of ribonuclease A by $Fe^{2+}+H_2O_2$ (Fenton's Reagent). *Biochim. Biophys. Acta*, **268**, 121–124.

Deutsch, J.C. (1998) Ascorbic acid oxidation by hydrogen peroxide. *Analyt. Biochem.*, **255**, 1–7.

Erbes, D.L., King, D. and Gibbs, M. (1979) Inactivation of hydrogenase in cell-free extracts and whole cells of *Chlamydomonas reinhardtii*. *Plant Physiol.*, **63**, 1138–1142.

Erin, A.N., Skrypin, V.V. and Kagan, V.E. (1985). Formation of α-tocopherol complexes with fatty acids. Nature of complexes. *Biochim. Biophys. Acta*, **815**, 209–214.

Esterbauer, H., Hayn, M., Abuja, P.M. and Claeyssens, M. (1991) Structure of cellulolytic enzymes. In: ACS Symp. Series 460 (*Enzymes in Biomass Conversion*). Ch. 23, pp. 301–312.

Evans, B.R., Margalit, R. and Woodward, J. (1993) Attachment of pentaammine ruthenium(III) to *Trichoderma reesei* cellobiohydrolase I increases its catalytic activity. *Biochem. Biophys. Res. Commun.*, **195**, 497–503.

Evans, B.R., Margalit, R. and Woodward, J. (1994) Veratryl alcohol oxidase activity of a chemically modified cellulase protein. *Arch. Biochem. Biophys.*, **312**, 459–466.

Evans, B.R., Margalit, R. and Woodward, J. (1995) Enhanced hydrolysis of soluble cellulosic substrates by a metallocellulase with veratryl alcohol oxidase activity. *Appl. Biochem. Biotechnol.*, **51/52**, 225–239.

Evans, R.M., Fraser, J.B. and Owen, L.N. (1949) Dithiols. Part III. Derivatives of polyhydric alcohols, *J. Chem. Soc.*, Part 1, 248.

Finlay, E.L., Busman, M., Dillon, J., Crouch, R.K. and Schey, K.L. (1997) Identification of photooxidation sites in bovine α-crystallin. *Photochem. Photobiol.*, **66**, 635–641.

Fischer, B., Sumner, I. and Goodenough, P. (1992) Isolation and renaturation of bio-active proteins expressed in *Escherichia coli* as inclusion bodies. *Arzneimittel-forschung*, **42**, 1512–1515.

Fokhina, N.A. and Velikodvorskaia, G.A. (1997) Cloning and expression of the gene for thermostable β-galactosidase from *Thermoanaerobacter ethanolicus* in *Escherichia coli*: purification and properties of the product. *Mol. Gen. Microbiol. Virol.*, **2**, 34–36.

Fuentes, A.M. and Amabile-Cuevas, C.F. (1998) Antioxidant vitamins C and E affect the superoxide-mediated induction of the soxRS regulon of *Escherichia coli*. *Microbiology*, **144**, 1731–1736.

Grassman, W., v. Schoenebeck, O. and Eibeler, H. (1931) Plant proteases (XVI). Activation of animal and plant proteases by glutathione. *Ztschr. Physiol. Chemie*, **194**, 124–136.

Gross, M.D., Bishop, T.D., Belcher, J.D. and Jacobs, D.R., Jr (1997) Solubilization of β-carotene in culture media. *Nutr. Cancer*, **27**, 174–176.

Guissani, A., Henry, Y. and Gilles, L. (1982) Radical scavenging and electron-transfer reactions in *Polyporus versicolor* laccase. A pulse radiolysis case study. *Biophys. Chem.*, **15**, 177–190.

Gutteridge, J.M. and Wilkins, S. (1983) Copper salt-dependent hydroxyl radical formation. Damage to proteins acting as antioxidants. *Biochim. Biophys. Acta*, **759**, 38–41.

Hahn, D. and Kück, U. (1994) Biochemical and molecular genetic basis of hydrogenases. *Process Biochem.*, **29**, 633–644.

Harbury, H.A. and Loach, P.A. (1960) Oxidation-linked proton functions in heme octa- and undecapeptides from mammalian cytochrome *c*. *J. Biol. Chem.*, **235**, 3640–3645.

Held, K.D., Sylvester, F.C., Hopcia, K.L. and Biaglow, J.E. (1996) Role of Fenton chemistry in thiol-induced toxicity and apoptosis. *Radiat. Res.*, **145**, 542–553.

Hershlag, N., Hurley, I. and Woodward, J. (1998) A simple method to demonstrate the enzymatic production of hydrogen from sugar. *J. Chem. Educ.*, **75**, 1270–1274.

Holland, H.D. (1984) *The Chemical Evolution of the Atmosphere and Oceans*. Princeton University Press, Princeton, NJ.

Holmgren, A. and Björnstedt, M. (1995) Thioredoxin and thioredoxin reductase. *Methods Enzymol.*, **252**, 199–208.

Hoey, B.M. and Butler, J. (1984) The repair of oxidised amino acids by antioxidants. *Biochim. Biophys. Acta*, **791**, 212–218.

Houghten, R.A. and Li, C.H. (1983) Reduction of sulfoxides in peptides and proteins. *Methods Enzymol.*, **91**, 549–559.

Hubbard, S.R. (1995) Characterization of the cysteine-rich zinc-binding domains of protein kinase C by x-ray absorption spectroscopy. *Methods Enzymol.*, **252**, 123–132.

Huber, C. and Wächterhäuser, G. (1997) Activated acetic acid by carbon fixation on (Fe, Ni)S under primordial conditions. *Science*, **276**, 245–247.

Ikeda, M. and Clark, D.S. (1998) Molecular cloning of extremely thermostable esterase gene from hyperthermophilic archaeon *Pyrococcus furiosus* in *Escherichia coli*. *Biotechnol. Bioeng.*, **57**, 624–629.

Jacob, R.A. and Burri, B.J. (1996) Oxidative damage and defense. *Am. J. Clin. Nutr.*, **63**, 985S–990S.

Jacobsen, K.B., Manos, R.E. and Wadzinski, F.A. (1987) Partial purification of an oxygen scavenging cell membrane fraction for use in anaerobic biochemical reactions. *Biotechnol. Appl. Biochem.*, **9**, 368–379.

Jialal, I., Norkus, E.P., Cristol, L. and Grundy, S.M. (1991) β-Carotene inhibits the oxidative modification of low-density lipoprotein. *Biochim. Biophys. Acta*, **1086**, 134–138.

Kaufman, P.B., Wu, W., Kim, D. and Cseke, L.J. (1995) Extraction and purification of proteins/enzymes. In: *Handbook of Molecular and Cellular Methods in Biology and Medicine*, Chapter 3. CRC Press, Boca Raton, FL, pp. 41–63.

Kennedy, T.A. and Liebler, D.C. (1992) Peroxyl radical scavenging by β-carotene in lipid bilayers. Effect of oxygen partial pressure. *J. Biol. Chem.*, **267**, 4658–4663.

Kengen, S.W.M., Luesink, E.J., Stams, A.J.M. and Zehnder, A.J.B. (1993) Purification and characterization of an extremely stable β-glucosidase from the hyperthermophilic archaeon *Pyrococcus furiosus*. *Eur. J. Biochem.*, **213**, 305–312.

Kido, K. and Kassel, B. (1975) Oxidation of methionine residues of porcine and bovine pepsins. *Biochemistry*, **14**, 631–635.

Klibanov, A.M., Kaplan, N.O. and Kamen, M.D. (1978) A rationale for stabilization of oxygen-labile enzymes: Application to a clostridial hydrogenase. *Proc. Natl. Acad. Sci. US*, **75**, 3640–3643.

Konz, J.O., King, J. and Cooney, C.L. (1998) Effects of oxygen on recombinant protein expression. *Biotechnol. Prog.*, **14**, 393–409.

Kraehenbuhl, J.P., Galardy, R.E. and Jamieson, J.D. (1974) Preparation and characterization of an immunoelectron microscope tracer consisting of a heme-octapeptide coupled to Fab. *FEBS Lett.*, **139**, 208–223.

Kremer, S.M. and Wood, P.M. (1992a) Evidence that cellobiose oxidase from *Phanerochaete chrysosporium* is primarily an Fe(III) reductase. Kinetic comparison with neutrophil NADPH oxidase and yeast flavocytochrome b2. *Eur. J. Biochem.*, **205**, 133–138.

Kremer, S.M. and Wood, P.M. (1992b) Production of Fenton's reagent by cellobiose oxidase from cellulolytic cultures of *Phanerochaete chrysosporium*. *Eur. J. Biochem.*, **208**, 807–814.

Laemmli, U.K. (1970) Cleavage of structural proteins during the assemblage of the head of bacteriophage T4. *Nature*, **227**, 680–685.

Lee, C.Y. and Zeikus, J.G. (1992) Purification and characterization of thermostable glucose isomerase from *Clostridium thermosulfurogenes* and *Thermoanaerobacter* strain B6A. *Biochem. J.*, **273**, 565–571.

Linder, M. and Teeri, T. (1997) The roles and function of cellulose-binding domains. *J. Biotechnol.*, **57**, 15–28.

Lindmark, D.G. and Müller, M. (1973) Hydrogenosome, a cytoplasmic organelle of the anaerobic flagellate *Tritrichomonas foetus*, and its role in pyruvate metabolism. *J. Biol. Chem.*, **248**, 7724–7728.

Liu, S.V., Zhou, J., Zhang, C., Cole, D.R., Gajdarziska-Josifovska, M. and Phelps, T.J. (1997) Thermophilic Fe(III)-reducing bacteria from the deep subsurface: the evolutionary implications. *Science*, **277**, 1106–1109.

Loach, P. (1968) Oxidation–reduction potentials, absorbance bands, and molar absorbance of compounds used in biochemical studies, In: *CRC Handbook of Biochemistry*, The Chemical Rubber Company, Cleveland, OH, H.A. Sober, R.A. Hart, and E.K. Sober eds, pp. J27–J34.

Mahan, L.C. and Insel, P.A. (1984) Use of superoxide dismutase and catalase to protect catecholamines from oxidation in tissue culture studies. *Anal. Biochem.*, **136**, 208–216.

Mao, G.D., Thomas, P.D., Lopaschuk, G.D. and Poznansky, M.J. (1993) Superoxide dismutase (SOD)-catalase conjugates. Role of hydrogen peroxide and the Fenton reaction in SOD toxicity. *J. Biol. Chem.*, **268**, 416–420.

Markey, P.E., Greenfield, P.F. and Kittrell, J.R. (1975) Immobilization of catalase and glucose oxidase on inorganic supports. *Biotechnol. Bioeng.*, **17**, 285–289.

Martinez, A., Olafsdottir, S., Haavik, J. and Flatmark, T. (1992). Inactivation of purified phenylalanine hydrolase by dithiothreitol. *Biochem. Biophys. Res. Commun.*, **182**, 92–98.

Mattioli, G.S. and Wood, B.J. (1986) Upper mantle oxygen fugacity recorded by spinel lherzolites. *Nature*, **322**, 626–628.

McCord, J.M. and Fridovich, I. (1968) The reduction of cytochrome *c* by milk xanthine oxidase. *J. Biol. Chem.*, **243**, 5753–5760.

McCord, J.M., Keeler, B.B. and Fridovich, I. (1971) An enzyme-based theory of obligate anaerobiosis: The physiological function of superoxide dismutase. *Proc. Natl. Acad. Sci. US*, **68**, 1024–1027.

Meinicke, A.R., Zavan, S.S., Ferreira, A.M., Vercesi, A.E. and Bechara, E.J. (1996) The calcium sensor ruthenium red can act as a Fenton-type reagent. *Arch. Biochim. Biophys.*, **328**, 239–244.

Meister, A. (1995) Glutathione metabolism. *Methods Enzymol.*, **251**, 3–7.

Miller, S.L. and Orgel, L.E. (1974) The primitive atmosphere, In: *The Origins of Life on the Earth* Chapter 4, Prentice-Hall, Englewood Cliffs, NJ, pp. 33–54.

Moroney, J.V., Andreo, C.S., Vallejos, R.H. and McCarty, R.E. (1980) Uncoupling and energy transfer inhibition of photophosphorylation by sulfhydryl reagents. *J. Biol. Chem.*, **255**, 6670–6674.

Mozhaev, V.V., Berezin, I.V. and Martinek, K. (1998) Structure–stability relationships in proteins: Tasks and strategies for the development of stabilized enzyme catalysts for biotechnology. *CRC Critical Rev. Biochem.*, **23**, 235–281.

Nakos, G. and Mortenson, L. (1971) Purification and properties of hydrogenase, an iron–sulfur protein, from *Clostridium pasteurianum* W5. *Biochim. Biophys. Acta*, **227**, 576–583.

Osman, A.M., Koerts, J., Boersma, M.G., Boeren, S., Veeger, C. and Rietjans, I.M. (1996) Microperoxidase/H_2O_2-catalyzed aromatic hydroxylation proceeds by a cytochrome-P-450-type oxygen transfer reaction mechanism. *Eur. J. Biochem.*, **240**, 232–238.

Ozhagina, O.A. and Kasaikina, O.T. (1995) β-Carotene as an interceptor of free radicals. *Free Radical Biol. Med.*, **19**, 575–581.

Pauling, L. (1970) *Vitamin C and the Common Cold.* W.H. Freeman, San Francisco.

Peck. H.D. and Gest, H. (1955) A new procedure for assay of bacterial hydrogenase. *J. Bacteriol.*, **71**, 70–80.

Perkins, R.C., Beth, A.H., Wilkerson, L.S., Serafin, W., Dalton, L.R., Park, C.R. and Park, J.H. (1980) Enhancement of free radical reduction by elevated concentrations of ascorbic acid in avian dystrophic muscle. *Proc. Natl. Acad. Sci. US*, **77**, 790–794.

Peters, J.W., Lanzilotta, W.N., Lemon, B.J. and Seefeldt, L.C. (1998) X-ray crystal structure of the Fe-only hydrogenase (CpI) from *Clostridium pasteurianum* to 1.8 Angstrom resolution. *Science*, **282**, 1853–1858.

Poole, D.M., Hazlewood, G.P., Huskisson, N.S., Virden, R. and Gilbert, H.J. (1993) The role of conserved tryptophan residues in the interaction of a bacterial cellulose binding domain with its ligand. *FEMS Microbiol. Lett.*, **106**, 77–84.

Price, D.L., Sisodia, S.S., Borchelt, D.R. (1998) Genetic neurodegenerative diseases: The human illness and transgenic models. *Science*, **282**, 1079–1083.

Prutz, W.A., Butler, J. and Land, E.J. (1996) Photocatalytic and free radical interactions of the heterocyclic N-oxide resazurin with NADH, GSH, and Dopa. *Arch. Biochim. Biophys.*, **327**, 239–248.

Raman, B., Ramakrishna, T. and Rao, C.M. (1996) Refolding of denatured and denatured/reduced lysozome at high concentrations. *J. Biol. Chem.*, **271**, 17067–17072.

Rao, K.K., Rosa, L. and Hall, D.O. (1976) Prolonged production of hydrogen gas by a chloroplast biocatalytic system. *Biochem. Biophys. Res. Commun.*, **68**, 21–28.

Robson, R.L. (1979) Characterization of an oxygen-stable nitrogenase complex isolated from *Azobacter chroococcum*. *Biochem. J.*, **181**, 569–575.

Rudd, P.M., Joao, H.C., Coghill, E., Fiten, P., Saunders, M.R., Opdenakker, G. and Dwek, R.A. (1994) Glycoforms modify the dynamic stability and functional activity of an enzyme. *Biochemistry*, **33**, 17–22.

Schneider, K., Cammack, R., Schlegel, H.G. and Hall, D.O. (1979) The iron–sulphur centres of soluble hydrogenase from *Alcaligenes eutrophus*. *Biochim. Biophys. Acta*, **578**, 445–461.

Shao, W. and Wiegel, J. (1992) Purification and characterization of a thermostable β-xylosidase from *Thermoanaerobacter ethanolicus*. *J. Bacteriol.*, **174**, 5848–5853.

Shintani, D. and DellaPenna, D. (1998) Elevating the vitamin E content of plants through metabolic engineering. *Science*, **282**, 2098–2100.

Shoulson, I. (1998) Experimental therapeutics of neurodegenerative disorders: unmet needs. *Science*, **282**, 1072–1074.

Simat, T., van Wickern, B., Eulitz, K. and Steinhart, E.H. (1996) Contaminants in biotechnologically manufactured L-tryptophan. *J. Chromatogr. B. Biomed. Appl.*, **685**, 41–51.

Sizer, I. (1946) The action of tyrosinase on proteins. *J. Biol. Chem.*, **163**, 145–157.

Sizer, I. (1947) The action of tyrosinase on certain proteins and products of their autolysis. *J. Biol. Chem*, **169**, 303–311.

Sizer, I. (1948) The inactivation of invertase by tyrosinase. *Science*, **108**, 335–336.

Stadtman, E.R. and Berlett, B.S. (1991) Fenton chemistry: Amino acid oxidation. *J. Biol. Chem.*, **266**, 17201–17211.

Stadtman, E.R. and Wittenberger, M.E. (1985) Inactivation of *Escherichia coli* glutamine synthetase by xanthine oxidase, nicotimate hydroxylase, horseradish peroxidase, or glucose oxidase: Effects of ferredoxin, putidaredoxin, and menadione. *Arch. Biochem. Biophys.*, **239**, 379–387.

Steinman, H.M., Vishweshar, R.N., Abernethy, J.L. and Hill, R.L. (1974) Bovine erythrocyte superoxide dismutase: Complete amino acid sequence. *J. Biol. Chem.*, **249**, 7326–7338.

Stern, M.H., Robeson, C.D., Weisler, L. and Baxter, J.G. (1947). δ-Tocopherol. I. Isolation from soybean oil and properties. *J. Am. Chem. Soc.*, **69**, 869–874.

Sunna, A., Puls, J. and Antranikian, G. (1996) Purification and characterization of two thermostable two endo-1,4-β-D-xylanases from *Thermotoga thermarum*. *Biotechnol. Appl. Biochem.*, **24**, 177–185.

Tien, M. and Kirk, T.K. (1984) Lignin-degrading enzyme from *Phanerochaete chrysosporium*: Purification, characterization and catalytic properties of a unique H_2O_2-requiring oxygenase. *Proc. Natl. Acad. Sci. US*, **81**, 2280–2284.

Tomme, P., Van Tilbeurgh, H., Pettersson, G., Van Damme, J., Vandekerckhove, J., Knowles, J., Teeri, T., Claeyssens, M. (1988) Studies of the cellulolytic system of *Trichoderma ressei* QM 9414: Analysis of domain function in two cellobiohydrolases by limited proteolysis. *Eur. J. Biochem.*, **170**, 575–581.

Torreilles, J. and Guerin, M.C. (1990) Nickel(II) as a temporary catalyst for hydroxyl radical generation. *FEBS Lett.*, **272**, 58–60.

Traverso, N., Menini, S., Cottalasso, D., Odetti, P., Marinari, U.M. and Pronzato, M.A. (1997) Mutual interaction between glycation and oxidation during non-enzymatic protein modification. *Biochim. Biophys. Acta*, **1336**, 409–418.

Tsuchihashi, H., Kigoshi, M., Iwatsuki, M. and Niki, E. (1995) Action of β-carotene as an antioxidant against lipid peroxidation. *Arch. Biochem. Biophys.* **323**, 137–147.

Tuisel, H., Sinclair, R., Bumpus, J.A., Asbaugh, W., Brock, B.J. and Aust, S.D. (1990) Lignin peroxidase H2 from *Phanerochaete chrysosporium*: Purification, characterization and stability to temperature and pH. *Arch. Biochem. Biophys.*, **279**, 158–166.

Tyurin, V.A., Carta, G., Tyruina, Y.Y., Banni, S., Day, B.W. Corongiu, F.P. and Kagan, V.E. (1997) Peroxidase-catalyzed oxidation of β-carotene in HL-60 cells and in model systems: Involvement of phenoxyl radicals. *Lipids*, **32**, 131–142.

Uchida, E., Morimoto, K., Kawasaki, N., Izaki, Y., Abdu Said, A. and Hayakawa, T. (1997) Effect of active oxygen radicals on protein and carbohydrate moieties of recombinant erythropoietin. *Free Radical Res.* **27**. 311–323.

Urey, H.C. (1952) On the early chemical history of the earth and the origin of life. *Proc. Natl. Acad. Sci. US*. **38**, 351–363.

Voorthorst, W.G.B. Eggen, R.I.L., Luesink, E.J. and deVos W.M. (1995) Characterization of the celB gene coding for β-glucosidase from the hyperthermophilic archaeon *Pyrococcus furiosus* and its expression and site-directed mutagenesis in *E. coli. J. Bacteriol.*, **177**, 7105–7111.

West, C.M. (1986) Current ideas on the significance of protein glycosylation. *Mol. Cell. Biol.*, **72**, 3–20.

Wiedau-Pazos, M., Goto, J.J., Rabizadek, S., Gralla, E.B., Roe, J.A., Lee, M.K., Valentine, J.S. and Bredesen, D.E. (1996) Altered reactivity of superoxide dismutase in familial amylotrophic lateral sclerosis. *Science*, **271**, 515–518

White, F.H. (1961) Regeneration of native secondary and tertiary structures by air oxidation of reduced ribonuclease. *J. Biol. Chem.*, **236**, 1353–1358.

Wiseman, A. and Woodward, J. (1974) Comparison of inactivation studies on yeast invertase by using mushroom tyrosinase and potassium nitrosyldisulphonate (Fremy's Salt). *Biochem. Soc. Trans.*, **2**, 594–596.

Wood, J.L. and duVigneaud, V. (1939) A new synthesis of cystine. *J. Biol. Chem.*, **131**, 267–271.

Woodward, F.N. (1948) Monothioethylene glycol and thiodiglycol. *J. Chem. Soc.*, 1892–1895.

Woodward, J., Mattingly, S.M., Danson, M., Hough, D., Ward, N. and Adams, M. (1996) In vitro hydrogen production by glucose dehydrogenase and hydrogenase. *Nature Biotechnol.*, **14**, 872–874.

Yan, L.-J., Orr, W.C. and Sohal, R.S. (1998) Identification of oxidised proteins based on sodium dodecyl sulfate-polyacrylamide gel electrophoresis, immunochemical detection, isoelectric focusing, and microsequencing. *Anal. Biochem.*, **263**, 67–71.

Young, A.J., Phillip, D. and Savill, J. (1997) Carotenoids in higher plant photosynthesis. In: *Handbook of Photosynthesis*, Chapter 3, M. Pessarakli, ed. Marcel Dekker, New York, pp. 575–596.

Zhang, C., Liu, S., Logan, J., Mazumder, R. and Phelps, T.J. (1996) Enhancement of Fe(III), Co(III), and Cr(VI) reduction at elevated temperatures and by a thermophilic bacterium. *Appl. Biochem. Biotechnol.*, **57/58**, 923–932.

Zhao, F., Ghezzo-Schoneich, E., Aced, G.I., Hong, J., Milby, T. and Schoneich, C. (1997) Metal-catalyzed oxidation of histidine in human growth hormone. Mechanism, isotope effects, and inhibition by a mild denaturing alcohol. *J. Biol. Chem.*, **272**, 9019–9029.

Zhao, F., Yang, J. and Schoneich, C. (1996) Effects of polyaminocarboxylate metal chelators on iron-thiolate induced oxidation of methionine and histidine-containing peptides. *Pharm. Res.*, **13**, 931–938.

Zidoni, E. and Kremer, M.L. (1974) Kinetics and mechanism of catalase action. Formation of the intermediate complex. *Arch. Biochem. Biophys.*, **161**, 658–664.

Gene Expression and Cell Mediated Mechanisms of Toxicity Expression

4 Xenobiotic Metabolism and Bioactivation by Cytochromes P-450

COSTAS IOANNIDES

Molecular Toxicology Group, School of Biological Sciences, University of Surrey, Guildford, Surrey, GU2 7XH, UK

There is no doubt that both the ever-increasing longevity and the high quality of life humans experience today could not have been achieved in the absence of the myriad of anthropogenic chemicals that form an integral part of modern life. It is not coincidental that the sharpest rise in life expectancy has occurred this century, which witnessed the birth of the pharmaceutical industry. The chemist has synthesised novel chemicals, not pre-existing in nature, which possess biological activity that can be exploited to treat effectively human disease and to provide a more hygienic environment.

Human exposure to a myriad of structurally diverse chemicals is continuous, from conception to death, and is unavoidable. It can be both voluntary, for example drugs and food additives, or involuntary, for example environmental and dietary contaminants. Although the term 'chemicals' is frequently taken to refer to anthropogenic chemicals, it must be emphasised that there are numerous naturally occurring chemicals, primarily of plant origin, to which humans are exposed on a daily basis. Such chemicals also possess biological activity, and during the last two decades it has become evident that they can afford protection to humans against major degenerative diseases such as cancer and cardiovascular disease (Mori and Nishikawa 1996; Ahmad *et al.* 1996), and may explain the epidemiological findings that populations consuming diets with high vegetable and fruit content are less susceptible to these fatal diseases.

The living organism can utilise some of these chemicals to generate energy (e.g. fats), as structural blocks or as essential cofactors for enzyme reactions. Most of the chemicals to which humans are exposed, however, cannot be used in the above processes and are often called 'xenobiotics' (Gr, foreign to life). The body recognises these compounds as being foreign and its first line of defence is to eliminate them. It is unlikely that the human body could withstand the constant onslaught of the chemicals to which it is unavoidably exposed if such defence mechanisms were not present. In order to gain entry in to the body, xenobiotics must traverse lipoid membranes, for example the gastrointestinal tract, and consequently are fairly lipophilic. Since lipophilic compounds that are excreted through the kidney by glomerular filtration and active secretion may be passively reabsorbed from the kidney tubule back into the blood circulation, and biliary excretion is available only for the excretion of hydrophilic compounds, the body must render these lipophilic compounds hydrophilic in order to eliminate them effectively. It has, therefore, developed a number of enzyme systems adept at metabolically converting lipophilic xenobiotics to hydrophilic metabolites. Moreover, such metabolism terminates the biological activity of chemicals since, in most cases, the metabolites, in contrast to the parent compounds, are unable to interact with receptors to elicit a biological effect. Xenobiotic metabolism occurs in two distinct phases, namely Phase I and Phase II.

Metabolism of xenobiotics

During Phase I metabolism, frequently referred to as functionalisation, an atom of oxygen is incorporated into the chemical, and functional groups such as –OH, –COOH, etc. are generated; alternatively, such functional groups may be unmasked for example an alkoxy (–C–OR) group can be dealkylated to unmask a functional group (–C–OH). The generated metabolites, not only are more polar than the parent compound but, moreover, can now undergo Phase II metabolism. The functional groups of Phase I metabolism are conjugated with polar endogenous chemicals such as sulphate, glucuronic acid and amino acids to form metabolites which, by virtue of their high polarity, are effectively eliminated from the body via the kidney and bile (Figure 4.1). Chemicals that already possess a group that can participate in conjugation reactions can bypass Phase I metabolism. Through this process of metabolism, the body eliminates xenobiotics that it cannot beneficially exploit.

Bioactivation of xenobiotics

A well-documented paradox of xenobiotic metabolism is that with certain chemicals metabolism, both Phase I and Phase II, can generate metabolites which are highly reactive. These metabolites can interact irreversibly with vital cellular constituents to provoke various types of toxicity; thus, in this case, metabolism confers adverse biological activity. The process through which inert chemicals are biotransformed to reactive intermediates capable of causing cellular damage is known as 'metabolic activation' or

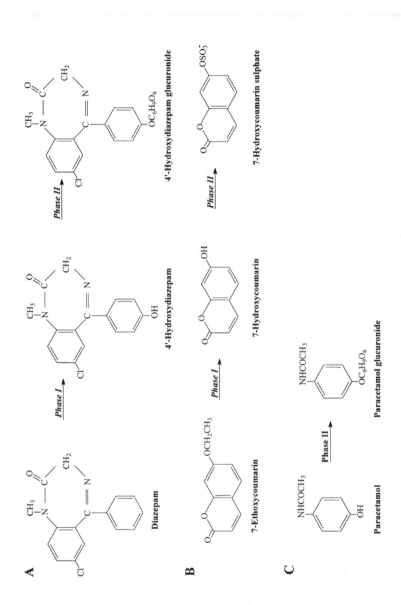

Figure 4.1. Metabolism of xenobiotics by Phase I and Phase II pathways. A. Generation of a functional group by Phase I metabolism and subsequent Phase II conjugation with glucuronic acid. B. Unmasking of a functional group by Phase I metabolism and subsequent Phase II conjugation with sulphate. C Direct Phase II conjugation with glucuronic acid

'bioactivation'. For such chemicals, carcinogenicity is inextricably linked to their metabolism. 4-Aminobiphenyl is an established human carcinogen inducing tumours in both urinary bladder and liver, yet it is chemically an inert compound. Enzyme systems in the body, however, can convert it to the more reactive hydroxylamine, which then conjugates with sulphate; the sulphate conjugate breaks down spontaneously to generate a highly reactive electrophile, the nitrenium ion that readily interacts covalently with DNA (Figure 4.2). The body attempts to repair the altered DNA, but if it fails to do so the damaged DNA can progress to a tumour. In this example, both Phase I and Phase II enzyme systems act in concert to form a deleterious metabolic product. The analgesic drug paracetamol (acetaminophen) undergoes oxidation to form a benzoquinoneimine capable of interacting with the –SH groups of proteins, this constituting an essential step of the mechanism leading to its hepatotoxicity and nephrotoxicity (Figure 4.3). The antimalarial drug amodiaquine, a 4-aminoquinoline, is also oxidised to a quinoneimine that functions as a hapten; it interacts with proteins to produce neoantigens that can provoke an immune response (Figure 4.4) (Jewell *et al.* 1995). Reactive intermediates can also interact with tissue oxygen to generate reactive oxygen species through redox cycling, giving rise to toxicity, including DNA injury. For example quinones, products of the oxidation of aromatic hydrocarbons, may be subject to one-electron flavoprotein reductions to yield the semiquinone radical. This can induce DNA damage directly, or may interact with molecular oxygen, producing reactive oxygen species, thus inducing DNA lesions indirectly (Figure 4.5). Clearly, toxicity is not simply a consequence of the intrinsic molecular structure of the chemical, but is determined by the nature of the enzymes present at the time of exposure; these enzyme systems are in turn regulated genetically but are also modulated by environmental factors such as diet and previous exposure to chemicals, presence of disease and physiological factors such as age and sex (see below).

The most effective protective mechanism against chemical reactive intermediates is their detoxification through conjugation with endogenous nucleophilic substrates such as the tripeptide glutathione, this being the second line of defence. This process not only results in the abolition of the biological activity of the reactive intermediate but also renders it sufficiently polar for its facile excretion. As already discussed (see above), paracetamol is activated to a reactive benzoquinoneimine; the body protects itself from its hepatotoxic effects by conjugating it with glutathione, thus preventing it from binding to liver proteins (Figure 4.3). Glutathione conjugates are excreted into the urine and bile usually following further processing. Clearly, a chemical is subject to a number of pathways, the majority of which will bring about its deactivation and facilitate its excretion. However, some routes of metabolism will transform the chemical to a metabolite

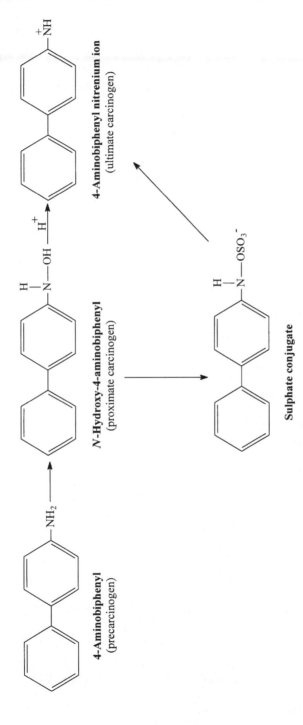

Figure 4.2. Bioactivation of 4-aminobiphenyl to a genotoxic species

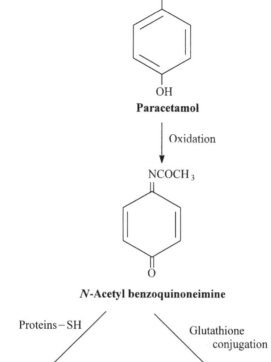

Figure 4.3. Bioactivation of paracetamol to a hepatotoxic species

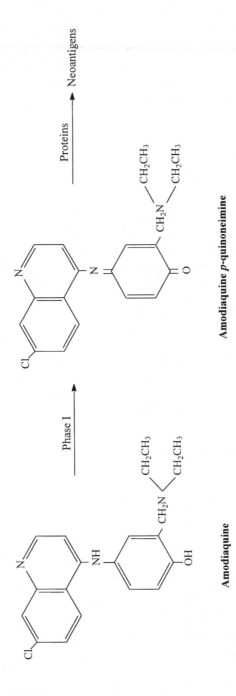

Figure 4.4. Bioactivation of amodiaquine to an immunotoxic species

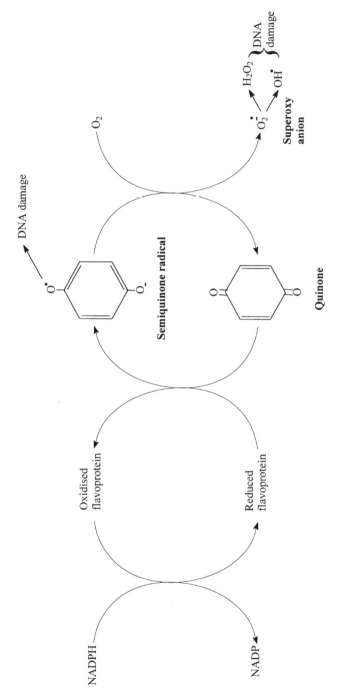

Figure 4.5.   Generation of reactive oxygen species from quinones

capable of inducing toxicity. Obviously, the amount of reactive intermediate produced, and hence incidence and degree of toxicity, will be largely dependent on the competing pathways of activation and deactivation. In most cases, the activation pathways represent minor routes of metabolism so that the generation of reactive intermediates is minimal, the low levels formed are effectively deactivated by the defensive mechanisms, and no toxicity is apparent. However, in certain situations, the activation pathways may assume a greater role, leading to enhanced production of reactive intermediates, overwhelming the deactivation pathways, resulting in their accumulation in the body, thus increasing the likelihood of interaction with cellular components with ensuing toxicity. Such situations may arise when:

1. Deactivation pathways are saturated. This can occur during exposure to large quantities of a chemical leading to depletion of the conjugating species or as a result of inadequate nutrition.
2. The enzyme systems catalysing the activation pathways are selectively induced as a result of prior exposure to other chemicals (Okey 1990) or the presence of disease (Ioannides et al. 1996).

It is evident that any factor that disturbs the delicate balance of activation/ deactivation will also influence the fate and toxicity of chemicals.

Chemical carcinogenesis

A plethora of structurally very diverse chemicals present in the human environment have been shown to induce tumours in rodents, and consequently have the potential to also cause cancer in humans. The vast majority of these carcinogens are inert chemicals, unable to interact with DNA to set into motion the processes leading to tumorigenesis. However, they can acquire the necessary reactivity through bioactivation to form reactive intermediates that manifest their carcinogenicity. These reactive intermediates are also subject to deactivation through conjugation with glutathione or hydration, and so the balance of activation/deactivation is crucial (see above). It is becoming increasingly evident that reactive intermediates can be transported from their site of generation to other tissues where they can interact with DNA, leading eventually to tumorigenesis, as shown in the case of polycyclic aromatic hydrocarbons and aromatic and heterocyclic amines (see above) (Wall et al. 1991; Kaderlik et al. 1994; Verna et al. 1996).

The bioactivation of chemical carcinogens may involve one or more metabolic steps catalysed by both Phase I and Phase II enzymes. The carcinogens may be first converted to the proximate carcinogen which, although more reactive than the parent compound, is not the species that interacts with DNA, for example the hydroxylamine of 4-aminobiphenyl. The entity that interacts with DNA, the ultimate carcinogen, is the nitrenium ion which is formed from the further metabolism of the proximate carcinogen (Figure 4.2, see above). Both the proximate and ultimate carcinogen can be metabolised through alternative routes to yield inactive, polar and readily excretable products. Thus the amount of ultimate carcinogen generated, and consequently the likelihood and extent of carcinogenic response, is determined by the net effect of the various activating/deactivating pathways. To illustrate this concept benzo[a]-pyrene, a ubiquitous carcinogenic polycyclic aromatic hydrocarbon, may be used as an example (Figure 4.6). The diol-epoxide (benzo[a]pyrene-7,8-dihydrodiol-9,10-epoxide) is considered to be the principal ultimate carcinogen, and three pathways are involved in its production. Initially, benzo[a]pyrene is oxidised to the primary 7,8-epoxide, which functions as the proximate carcinogen. This is detoxicated through two different pathways, firstly conjugation with glutathione, catalysed by the glutathione S-transferases, and secondly through hydration, catalysed by the epoxide hydrolases, converting the epoxide to the corresponding dihydrodiol. The latter can be excreted following conjugation with sulphate or glucuronide but, in addition, it may be subjected to a second oxidation to yield the diol-epoxide, the ultimate carcinogen (Figure 4.6). The diol-epoxide is not a substrate of the epoxide hydrolases, and only a poor substrate of the glutathione S-transferases.

The susceptibility of an animal species to a given chemical carcinogen is largely dependent on its complement of enzymes at the time of exposure, which determines whether activation or deactivation of pathways are favoured. For example the guinea-pig is unable to carry out the N-hydroxylation of 2-acetylaminofluorene, the initial and rate-limiting step in its bioactivation and is consequently resistant to its carcinogenicity (Kawajiri et al. 1978). Similarly, the balance of activation/deactivation may determine which tissues within an animal species develop tumours. Aromatic amines are readily activated in the liver and are consequently potent hepatocarcinogens. However, many tissues with very limited activation ability also frequently develop tumours, indicating that reactive intermediates may be transported from their tissue of origin to other distant tissues. For example, when the liver from rats pretreated with benzo[a]-pyrene were transplanted in untreated animals, the extent of DNA binding in the lung, liver and kidney was the same in both, those exposed directly to the carcinogen and those who received the liver transplants, indicating clearly that the liver was the source of the reactive intermediates that interacted with DNA, not only for the liver but also for the lung and kidney (Wall et al. 1991). Similarly, the aromatic amines are potent urinary bladder

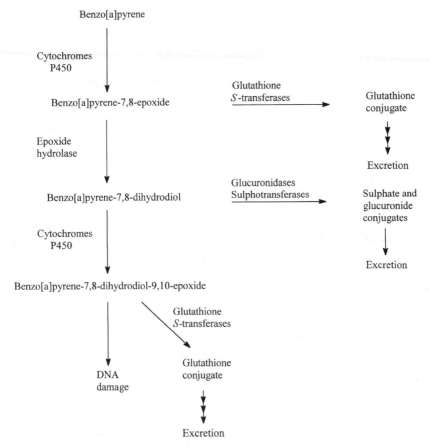

Figure 4.6.   Principal pathways of bioactivation and deactivation of benzo[a]pyrene

carcinogens despite the fact that the ability of the bladder to activate them metabolically is poor; it appears that reactive intermediates can be transported from the liver to this tissue where they can exert their carcinogenic effect (Kadlubar *et al.* 1981). Similarly, the breast is a frequent site of tumorigenesis, despite the fact that its capacity to bioactivate chemicals is poor (Davis *et al.* 1994).

Cytochromes P-450

Although a number of enzyme systems can catalyse the Phase I metabolism of xenobiotics, both activation and deactivation, undoubtedly the most important are the cytochrome P-450-dependent mixed function oxidases. This is a ubiquitous enzyme system being found in both prokaryotic and eukaryotic cells. In mammals they are membrane bound and, with the exception of striated muscle and red blood cells, are encountered in every tissue, and predominate in the liver which, as a consequence, is the principal site of xenobiotic transformations. However, cytochrome P-450 proteins are also active in portals of chemical entry such as the lungs, gastrointestinal tract and nasal mucosa. In the presence of O_2 and NADPH, it catalyses the incorporation of one atom of oxygen into the substrate while the second atom is converted into water, the haemoprotein cytochrome P-450 acting as the terminal oxidase (Porter and Coon 1991; Guengerich 1993). Almost every xenobiotic that finds its way into the human body is to some extent, at least, metabolised by cytochrome P-450 enzymes. The vast majority of eukaryotic cytochromes P-450 are localised in the endoplasmic reticulum (microsomes), but a small number of specialised forms active in steroid metabolism can be found in the inner mitochondrial membrane. Cytochromes P-450 have a monomeric molecular weight of about 50 000 and are believed to have evolved from a common ancestor some 3000 million years ago (Lewis *et al.* 1998). The function of the earliest forms is considered to have been in the metabolism (both biosynthesis and degradation) of essential endogenous chemicals, such as steroid hormones, and then evolved to enzymes capable of facilitating the elimination of xenobiotics. As a result, the cytochrome P-450 proteins lost their narrow substrate specificity towards steroids and, in order to cope with the new, increasingly chemical environment, they developed into broad specificity enzymes that could metabolise, and thus facilitate the elimination, of xenobiotics. It has been proposed that what forced the evolution of these proteins is the necessity to develop defence mechanisms to protect against plant toxins, present in the food chain, that have been produced in order to discourage predators (Gonzalez and Nebert 1990; Gonzalez and Gelboin 1992).

A major functional characteristic of the cytochrome P-450 system is its unprecedented versatility in catalysing effectively the metabolism of structurally very diverse xenobiotics. Its substrates range from small molecules such as methanol (MW=42) to large molecules such as the immunosuppressant drug cyclosporin (MW=1203). A single enzyme cannot accommodate such a variety of substrates, and indeed the cytochrome P-450 system achieves this very broad specificity by existing as a superfamily of enzymes. The cytochrome P-450 (CYP) superfamily is divided into families, denoted by an arabic number, which in turn are further subdivided into subfamilies, denoted by a capital letter, that may comprise one or more individual enzymes (isoforms), denoted by an arabic number. For example CYP3A4 represents an enzyme belonging to family 3, subfamily A, protein

4. The criteria for this classification are strictly structural with no consideration of their substrate specificity. Proteins within a family display at least a 40% structural homology, whereas between proteins within the same subfamily homology is at least 55%.

The various cytochrome P-450 proteins possess different, albeit overlapping, substrate specificity. They also display regioselectivity in that they may metabolise the same substrate but at different site (Oguri *et al.* 1994). For example, CYP2A1 hydroxylates testosterone at the 7α-position, whereas CYP3A4 is the principal catalyst of the 6β-hydroxylation (Chang and Waxman 1996; Maurel 1996). Similarly, the carcinogen β-naphthylamine is metabolised by CYP1A2 through N-hydroxylation, an activation step, whereas CYP1A1, and 2B catalyse its ring-oxidation at positions 1-and/or 6-, both being deactivation reactions (Hammons *et al.* 1985). More than one cytochrome P-450 enzyme may catalyse the same pathway of metabolism of a substrate but with different K_m values. For example, in humans planar aminocompounds are activated through N-hydroxylation catalysed principally by CYP1A2 and to a lesser extent by CYP1A1. However, the carcinogenic aromatic amine 2-amino-3,4-dimethylimidazo[4,5-f]quinoline, a dietary carcinogen, is also activated by CYP3A4/A5, but with rates being orders of magnitude lower than CYP1A2 (McManus *et al.* 1990). At the low concentrations that are likely to be achieved in the human liver, it is unlikely that the CYP3A proteins will be important catalysts. However, some contribution cannot be ruled out, as the CYP3A proteins are the most abundant cytochrome P-450 isoforms in the human liver (Wrighton and Stevens 1992; Shimada *et al.* 1997). Site-directed mutagenesis studies have revealed that small structural changes of even a single amino acid in the sequence can lead to marked changes in the substrate specificity and catalytic activity of individual cytochrome P-450 isoforms (Lindberg and Negishi 1989). The expression of individual cytochrome P-450 proteins may be monitored using diagnostic substrates selective for the isoform in question, and immunologically using appropriate antibodies.

Cytochrome P-450 families may be broadly classified into two categories. Those involved almost exclusively in the metabolism of endobiotic substrates (Table 4.1), with negligible catalytic activity towards xenobiotics, and those whose primary function is the metabolism of xenobiotics. The cytochrome P-450 proteins responsible for the metabolism of endogenous compounds are characterised by narrow substrate specificity, frequently entailing a single, or very few, structurally related substrates. The cytochrome P-450 families active in xenobiotic metabolism are CYP1 to CYP3; although they may catalyse the metabolism of endogenous substrates, such as steroids, they do so poorly. The fact that some xenobiotic-metabolising cytochrome P-450 isoforms are polymorphically expressed (see below), and individuals

Table 4.1. Endogenous substrates of cytochrome P-450

Class	Example
Steroids	Testosterone, progesterone
Sterols	Cholesterol, lanosterol
Fatty acids	Lauric acid, myristic acid
Prostaglandins	Prostglandin A_1 and E_1
Leukotrienes	Leukotriene B_4
Vitamins	Vitamin A, vitamin D_3
Porphyrinogens	Uroporphyrinogen

may totally lack a certain isoform without any detriment to health, indicates that the role of these cytochromes P-450 to endogenous metabolism is far from critical.

Cytochrome P-450 proteins can also function as generators of deleterious reactive oxygen species when NADPH is oxidised in the absence of substrate metabolism (Ingelman-Sundberg and Johansson 1984; Parke et al. 1991; Puntarulo and Cederbaum 1998). Superoxide can be released which, by the action of superoxide dismutase, may be converted to hydrogen peroxide which, in the presence of iron, is converted to the deleterious hydroxyl radical ($^\bullet$OH).

Induction and inhibition of cytochromes P-450

The activity of cytochromes P-450 is regulated by many factors including sex (Mugford and Kedderis 1998), presence of disease (Ioannides et al. 1996), diet and nutrition (Ioannides 1998) and levels of circulating hormones (Westin et al. 1992; Shapiro et al. 1995). One of the earliest observations on the regulation of cytochrome P-450 expression was its modulation by exposure to chemicals (Estabrook 1996). Extensive animal studies revealed that numerous chemicals, including many drugs such as barbiturates, isoniazid, hypolipidaemics, antiinflammatory, etc. can selectively upregulate the expression of cytochrome P-450 isoforms. Such upregulation of cytochrome P-450 expression has also been observed in human tissues following exposure to alcohol, smoking, polyhalogenated biphenyls, barbiturates, dietary indoles, etc. Most inducing agents are substrates of the cytochrome P-450 isoform they induce, i.e. they induce their own metabolism.

It is now recognised that xenobiotic-metabolising cytochrome P-450 enzymes fall into two categories: those expressed in untreated animals and which are resistant to induction (constitutive), and those that are poorly or not at all expressed constitutively but are readily inducible. The former group includes proteins belonging to the CYP2D subfamily and the latter includes all the proteins within the CYP1 family.

The molecular mechanisms of enzyme induction are currently being unravelled and appear to involve both transcriptional and post-transcriptional effects; it appears that the induction process varies among the cytochrome P-450 proteins. Induction of the CYP1 family has been extensively studied and is the best understood, following the pioneer work of Nebert (Nebert, 1989). A receptor protein residing in the cytosol, the Ah (aryl hydrocarbon) receptor, has been identified. This exists as a dimer with a heat-shock protein (hsp90) which is released following interaction with the inducing agent. The inducer/receptor translocates into the nucleus where it binds with another protein, the aryl hydrocarbon nuclear transferase, and the complex interacts with a specific region of the CYP1A1 gene, the xenobiotic regulatory element. This interaction leads to enhanced transcription of CYP1A1 and of other enzymes associated with this receptor (Hankinson 1995).

Exposure to chemicals can also lead to downregulation of cytochrome P-450 enzymes, an effect that is more frequent in clinical therapeutics than upregulation. The most common mechanism of inhibition of cytochrome P-450 activities involves competition by two or more substances for the substrate-binding site of the same cytochrome P-450 isoform; such chemicals rely on the same cytochrome P-450 enzyme for their metabolism. Another, reversible form of inhibition involves the interaction of a chemical, or its metabolite, with the haem iron of the cytochrome, for example antifungal drugs such as ketoconazole, the H_2-receptor antagonist cimetidine macrolide antibiotics, amphetamines and the naturally occurring methylene dioxyphenyls such as safrole, the last three examples necessitating prior cytochrome P-450-mediated metabolism to yield the interacting metabolite (Figure 4.7) (Knodell et al. 1991; Ioannides et al. 1981). Certain chemicals possessing olefinic and acetylenic groups can be metabolised to reactive intermediates, such as epoxides, which can destroy the cytochrome. Such inhibitors are drugs like secobarbital (Murray 1992; Murray and Reidy 1990) and naturally occurring chemicals like diallyl sulphide, a major constituent of garlic (Jin and Baillie 1997). An interaction with clinical implications between a dietary component and drugs has recently been described; consumption of a single glass of normal strength of grapefruit juice led to higher plasma levels, and consequently more pronounced pharmacological effect than anticipated, of drugs like felodipine, quinidine and midazolam (Spence 1997; Bailey et al. 1998). The mechanism involves inhibition of intestinal CYP3A4 activity by a component of grapefruit juice, believed to be a furanocoumarin, leading to impairment of the presystemic metabolism of these drugs.

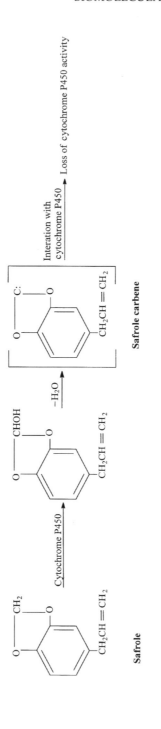

Figure 4.7. Metabolism of safrole to a cytochrome P450-binding species

Table 4.2.   Human and rat xenobiotic-metabolising cytochrome P-450 isoforms

Cytochrome P-450 subfamily	Human isoforms	Rat isoforms
CYP1A	1A1, 1A2	1A1, 1A2
CYP1B	1B1	1B1
CYP2A	2A6, 2A7, 2A13	2A1, 2A2, 2A3
CYP2B	2B6	2B1, 2B2, 2B3, 2B12, 2B15, 2B16
CYP2C	2C8, 2C9, 2C18, 2C19	2C6, 2C7, 2C11, 2C12, 2C13, 2C22, 2C23, 2C24
CYP2D	2D6	2D1, 2D2, 2D3, 2D4, 2D5, 2D18
CYP2E	2E1	2E1
CYP2F	CYP2F1	
CYP2G		CYP2G1
CYP3A	3A4, 3A5, 3A7	3A1, 3A2, 3A9, 3A18, 3A23
CYP4A	4A9, 4A11	4A1, 4A2, 4A3, 4A8
CYP4B	CYP4B1	

Genetic polymorphism in cytochrome P-450 expression

Genetic polymorphism in cytochrome P-450 activity was first recognised with respect to the human CYP2D6 protein, an enzyme active in the metabolism of many drugs (Gonzalez 1996; Rendic and Di Carlo 1997). About 5–10% of Caucasians lack a functionally active protein so that they are susceptible to drugs which rely largely on this protein for their metabolism, and such individuals are termed 'poor metabolisers'. Subsequently, other polymorphically expressed cytochrome P-450 proteins have been identified belonging to the CYP2C and CYP1A subfamilies (Richardson and Johnson 1996; Eaton et al. 1995).

The major characteristics of each of the xenobiotic-metabolising families will be discussed with emphasis on their role in the bioactivation of chemical carcinogens. Much of the information is derived from studies employing cytochrome P-450 proteins which, through recombinant DNA technology, have been expressed in bacteria and other cells (Guengerich et al. 1996). Table 4.2 illustrates the rat and human cytochrome P-450 proteins involved in xenobiotic metabolism.

Cytochrome P-450-dependent metabolism of drugs and bioactivation of chemical carcinogens

The CYP1 family

Two subfamilies make up the CYP1 family, 1A and 1B; the former comprises two proteins, 1A1 and 1A2, whereas only a single protein belonging to the 1B subfamily has so far been identified. CYP1A2 is almost exclusively a hepatic protein and appears not to be expressed at the protein level in most extrahepatic tissues, whereas CYP1A1 is largely expressed in extrahepatic tissues and is readily inducible in the liver (Guengerich 1990; Eaton *et al*. 1995). The CYP1B1 isoform is distributed in many tissues, and high constitutive expression was noted in the adrenals (Walker *et al*. 1995). The CYP1 family, originally known as cytochrome P-448, was one of the first to be identified and consequently it has been extensively studied.

The substrates of the CYP1 family are essentially lipophilic planar compounds characterised by a small depth and a large area/depth^2 ratio (Lewis *et al*. 1986, 1987). 4-Aminobiphenyl is such a planar molecule and is readily *N*-hydroxylated by CYP1A, but its isomer, 2-aminobiphenyl, by virtue of the amino group being at the *ortho*-position is non-planar and, consequently, not similarly *N*-hydroxylated by CYP1A (Ioannides *et al*. 1989). However, the width of the molecule is also critical as molecules characterised by a large depth may be denied access to the substrate-binding site (Lewis *et al*. 1994). Computer modelling studies revealed that the molecular dimensions of a putative access channel of the CYP1A1 isoform has a depth of 3.261 Å, hence the requirement of planarity, and a width of 8.321 Å, so that molecules having a larger width, even if planar, are excluded, for example benzo(e)pyrene and 4-acetylaminofluorene (Lewis *et al*. 1994). It plays a limited role in the metabolism of drugs (Eaton *et al*. 1995), but it is the principal catalyst of drugs like theophylline, caffeine and phenacetin, and contributes significantly to the metabolism of drugs such as propranolol, imipramine and dantrolene. In contrast, it is the most important family in the bioactivation of chemicals including many major and ubiquitous classes of carcinogens to which humans are involuntarily and unavoidably exposed to, such as polycyclic aromatic hydrocarbons, heterocyclic and aromatic amines, mycotoxins, as well as of drugs like paracetamol (acetaminophen), dihydralazine and diethylstilboestrol and toxins like ipomeanol (Table 4.3) (Steele *et al*. 1983; Roy *et al*. 1992; Gonzalez and Gelboin 1994; Guengerich and Shimada 1991; Czerwinski *et al*. 1991; Ioannides and Parke 1990). It has also been shown to contribute significantly to the activation of the tobacco-specific nitrosamine 4-(methylnitrosamino)-1-(3-pyridyl)-1-butanone (NNK) (Smith *et al*. 1992; Crespi *et al*. 1991). They are believed to be responsible for the activation of more than 90% of known carcinogenic chemicals (Rendic and Di Carlo 1997). CYP1A2 has been shown to also be the major catalyst of the 2-hydroxylation of oestrogen (Aoyama *et al*. 1990a).

Most ubiquitous environmental and dietary carcinogens to which humans are frequently exposed are planar in nature, and therefore favoured substrates of the CYP1 family. The CYP1 family through arene oxidation and *N*-oxidation activates polycyclic aromatic hydrocarbons,

Table 4.3.   Chemicals activated by CYP1

Class	Example
Polycyclic aromatic hydrocarbons	Benzo[a]pyrene
Aromatic amines	4-Aminobiphenyl
Heterocyclic amines	2-Amino-3-methylimidazo-[4,5-f]quinoline (IQ)
Aromatic amides	2-Acedtylaminofluorene
Azocompounds	Dimethylaminoazobenzene
Polyhalogenated biphenyls	3,4,3',4'-Tetrachlorobiphenyl
Nitrosamines	4-(Methylnitrosamino)-1-(3-pyridyl)-1-butanone (NNK)
Mycotoxins	Aflatoxin B_1
Furans	Ipomeanol
Certain drugs	Paracetamol
Steroids	Oestradiol
Synthetic steroids	Diethylstilboestrol

aromatic amines, aflatoxins and heterocyclic amines. Generally, CYP1A1 is more effective in catalysing arene oxidation and CYP1A2 N-oxidation (Iannides and Parke 1990, 1993). The latter isoform has also been shown to play a prominent role in the activation of the mycotoxin aflatoxin B_1 through arene oxidation (Aoyama et al. 1990b). CYP1A2 is also responsible for the metabolism of dihydralazine to form protein-binding metabolites, binding also to the enzyme itself, thus forming a neoantigen being responsible for the hepatitis associated with this drug (Beaune and Bourdi 1993). Both CYP1A proteins, especially CYP1A1, are involved in the hydroperoxide-dependent oxidation of catechol oestrogens including the carcinogenic synthetic stilbene, diethylstilboestrol, to the genotoxic quinones (Roy et al. 1992). Recent studies have shown that CYP1B1 is also effective in catalysing the metabolism and activation, through arene oxidation, of polycyclic aromatic hydrocarbons such as benzo[a]pyrene, but less effectively compared than CYP1A1; it is also capable of activating aromatic amines (Shimada et al. 1996; Kim et al. 1998). CYP1B1 also hydroxylates oestradiol, primarily at the 4-position, whereas CYP1A1 metabolises the steroid less efficiently, at different positions (Hayes et al. 1996). In more recent studies, CYP1A2 was shown to be a major catalyst of the 2- and 4-hydroxylations of oestradiol and oestrone (Yamazaki et al. 1998). It is noteworthy that CYP3A proteins (see below) also catalyse these hydroxylations and are likely to play a prominent role in humans displaying high hepatic CYP3A4 activity (Yamazaki et al. 1998; Shou et al. 1997). It has been suggested that the 4-hydroxyoestradiol can directly interact covalently with DNA, or be a precursor for such a genotoxic metabolite, probably a quinone or a semiquinone which may also act as a generator of reactive oxygen species (Service 1998).

The CYP1 is probably the most conserved family within the phylogenetic tree so that the human proteins share extensive structural similarity and display similar specificity to the orthologous rodent proteins, alluding to an important role for this family in a vital biological function (Kawajiri and Hayashi 1996). A role has been ascribed for this family in the catabolism and elimination of bilirubin (De Matteis *et al.* 1991). Subsequent studies employing purified cytochrome P-450 proteins in reconstituted systems established that CYP1A2 catalyses the oxidation of uroporphyrinogen to uroporphyrin (Lambrecht *et al.* 1992). The fact that the levels of CYP1 are downregulated in the adult animal implies an important role in foetal and neonatal development (Lum *et al.* 1985). Because of its prominent role in the activation of chemical carcinogens, high levels of CYP1 are considered undesirable (Ioannides and Parke 1993).

The CYP1 is not a major family in the liver of animals and humans, comprising less than 5% of the cytochrome P-450 content in the liver of rats and about 10% in human liver. However, it is probably the most inducible family, being induced by planar compounds. Studies in rats have shown that a single intraperitoneal injection of as little as 50 μg/kg of benzo[a]pyrene, a polycyclic aromatic hydrocarbon, doubled the hepatic activity (Phillipson *et al.* 1984). Potent inducing agents such as 3-methylcholanthrene may upregulate the CYP1 family so that it comprises as much as 80% of the total hepatic cytochrome P-450 content (Ryan and Levin 1990). Induction is not confined to the liver, but in humans lung and placental activity was also induced as a result of smoking (Sesardic *et al.* 1988; Petruzzelli *et al.* 1988; Pasanen *et al.* 1990). Changes in the human diet, such as high intake of cruciferous vegetables, resulted in increased activity that led to lower plasma levels of drugs such as phenacetin (Conney *et al.* 1977; Kall *et al.* 1996). This effect has been attributed to the presence of indoles, particularly indole-3-carbinol. Numerous other dietary anutrients, including compounds formed by the cooking process, have been shown to upregulate CYP1 activity in the liver and intestine of animals (Ioannides 1999). Moreover, CYP1A2, but not CYP1A1, is upregulated by ingestion of corn oil and medium chain triglycerides, but both isoforms are downregulated by vitamin C deficiency (Ioannides 1998). However, starvation increases the expression of CYP1A1 in the liver but, in contrast, decreases the expression of CYP1A2. In humans CYP1 expression can also be upregulated, in both the liver and intestine, by exposure to therapeutic doses of drugs such as omeprazole (Diaz *et al.* 1990; McDonnell *et al.* 1992).

CYP1 activity has been closely linked to human cancer incidence. High levels of CYP1A in human peripheral lymphocytes have been directly correlated with susceptibility to lung cancer in smokers (Kellermann *et al.* 1973). These findings were questioned for some time, but subsequent studies employing improved methodology confirmed the findings (Kouri *et al* 1982; Korsgaard *et al.* 1984). The belief was that humans displaying high levels of CYP1 as a result of smoking were more effective in metabolically converting polycyclic aromatic hydrocarbons to their genotoxic metabolites. Good positive relationships have been established between lung cancer and CYP1A1 in the lung of smokers (Anttila *et al.* 1991, 1992). What is of interest is that among smokers those who contract cancer display high CYP1A1 activity (Bartsch *et al.* 1992). It is noteworthy that the bioactivation of polycyclic aromatic hydrocarbons, carcinogenic components of tobacco smoke, by CYP1A appears to occur in the same part of the airways and in the same cell types in which peripheral carcinomas are observed (Anttila *et al.* 1992). However, it is still not clear whether the increase in CYP1A is an important predisposing factor in carcinogenesis or is merely a consequence of the disease. Finally, CYP1A activity, as measured by the aryl hydrocarbon hydroxylase (AHH) activity, has been related inversely to survival in human breast cancer; patients with high AHH activity were associated with poor prognosis (Pykko *et al.* 1991).

The CYP2 family

This is one of the largest cytochrome P-450 families comprising a number of distinct subfamilies.

The CYP2A subfamily

At least three isoforms appear to be present in the rat, CYP2A1, 2A2 and 2A3, and the same number has been identified in humans, CYP2A6, 2A7 and 2A13. A characteristic of this subfamily is that the rodent proteins exhibit different substrate specificity when compared with the human orthologues. The rat protein CYP2A1 effectively hydroxylates steroids at the 7α-position, but is unable to hydroxylate the anticoagulant drug coumarin at the 7-position, whereas the human CYP2A6 readily hydroxylates coumarin but is not a steroid hydroxylator (Waxman *et al.* 1991). Rat CYP2A2, structurally similar to CYP2A1, hydroxylates testosterone principally at the 15α-position (Aoyama *et al.* 1990a). A characteristic of the CYP2A1 enzyme is its expression in the Leydig cells of the testes (Sonderfan *et al.* 1989). The last member of this subfamily in the rat, CYP2A3, appears to be expressed in the lung and olfactory mucosa, but not in the liver or other tissues (Kimura *et al.* 1989; Liu *et al.* 1996). The human proteins are CYP2A6 and the related CYP2A7, but these comprise only about 1% of the total hepatic cytochrome P-450; the former isoform has also been detected in human olfactory mucosa (Thornton-Manning *et al.* 1997). It plays only a modest role in the metabolism of drugs and bioactivation of xenobiotics. CYP2A6 contributes to the 4-hydroxylation of the anticancer drug cyclophosphamide, leading to the formation of the biologically active acrolein and phosphoramide mustard; similarly it catalyses the 4-hydroxylation of ifosphamide (Chang and Waxman 1996). In rats hepatic CYP2A1 is downregulated by choline deficiency, but upregulated by vitamin A supplementation and starvation (Ioannides 1998).

The role of CYP2A proteins in the bioactivation of environmental chemical carcinogens is limited. They appear to make a significant contribution to the human hepatic metabolic activation of N-nitrosamines, including the tobacco nitrosamines, and of 1,3-butadiene (Yamazaki et al. 1992; Crespi et al. 1990; Duescher and Elfarra 1994); however, the principal catalyst of the bioactivation of these compounds is CYP2E1 (see below). A role has also been ascribed for this subfamily in the activation of the fungal contaminant aflatoxin B_1 (Aoyama et al. 1990b); furthermore it may make a small contribution to the activation, through N-hydroxylation, of the carcinogen 4,4-methylenebis(2-chloroaniline) (MOCA) (Guengerich and Shimada 1991). The olfactory forms CYP2A3 in rats and CYP2A6 in humans can bioactivate the nasal toxin 2,6-dibenzonitrile and the carcinogen hexamethylphosphoramide (Liu et al. 1996).

In rats, CYP2A1 is modestly induced by chemicals such as phenobarbital and β-naphthoflavone which, however, are much more potent inducers of the CYP2B and CYP1A subfamilies respectively (Honkakoski and Negishi 1997). In contrast, the same inducing agents downregulate the hepatic levels of CYP2A2. The anticancer drug cisplatin has also been shown to stimulate the expression of CYP2A1 in the liver of rats (Chang and Waxman 1996). CYP2A3 also appears to be inducible (Kimura et al. 1989). There is no evidence that the human CYP2A proteins are inducible. Marked interindividual variation in CYP2A6 expression in human liver has been observed (Pelkonen and Raunio 1995).

The CYP2B subfamily

In rats the CYP2B subfamily comprises at least six proteins, B1 and B2 being the most extensively studied since they are the major isoforms induced by barbiturates such as phenobarbitone, the first recognised inducer of the cytochrome P-450 system. In the liver of humans it is a minor subfamily, representing less than 2% of the total cytochrome P-450 content.

CYP2B displays a very broad substrate specificity, metabolising a wide variety of chemicals (Nims and Lubet 1996). CYP2B1 and B2 appear to have identical substrate specificity, but the former displays higher catalytic activity. Among drugs, they are the major catalysts of the bioactivation of cyclophosphamide to its pharmacologically active metabolites (see above), and of cocaine in the rat (Chang *et al.* 1993; Poet *et al.* 1994). CYP2B is also active in the metabolism of drugs such as benzphetamine, hexobarbital and diazepam (Rendic and Di Carlo 1997). Generally, the role of the CYP2B subfamily in the metabolism of drugs in humans is minor.

It is involved in the metabolism of many chemical carcinogens, but appears to direct the metabolism primarily through the formation of inactive metabolites, so that a deactivating role has been ascribed to them (Parke and Ioannides 1990). For example, CYP2B proteins can metabolise aromatic amines and amides effectively, but they are unable to carry out N-hydroxylation, the activation pathway, and only catalyse ring oxidations that lead to biologically inactive metabolites (Astrom and De Pierre 1985; Hammons *et al.* 1985). However, they participate in the bioactivation of a number of long-chain nitrosamines to genotoxic metabolites (Shu and Hollenberg 1997). The human CYP2B6 protein has been shown recently to contribute to the activation of cyclophosphamide, aflatoxin B_1 and NNK (Code *et al.* 1997).

The CYP2B subfamily is highly inducible in rats by numerous structurally diverse chemicals including drugs like phenobarbital and diphenylhydantoin, pesticides like DDT and environmental contaminants such as non-planar polyhalogenated biphenyls (Nims and Lubet 1996). Its levels are also induced by many dietary anutrients including indole-3-carbinol, phenethyl isothiocyanate, diallyl sulphide, flavonoids, terpenoids, safrole and alcohol (Ioannides 1999). Enhanced levels have also been observed in the liver of animals exposed to corn oil or medium chain triglycerides (Ioannides 1998). Interestingly, starvation upregulates CYP2B1 but downregulates CYP2B2 (Imaoka *et al.* 1990).

The CYP2C subfamily

This is a large subfamily playing an important role in the metabolism of many clinically important drugs. It is polymorphically expressed, resulting in slow and fast metabolisers of its drug substrates (Goldstein and de Morais 1994).

It is responsible for the metabolism of many major drugs such as hydantoins, diazepam, tolbutamide, warfarin, taxol, imipramine, hexobarbital, omeprazole, propranolol and non-steroidal anti-inflammatory drugs including ibuprofen and diclofenac. It is of interest that many of these drugs whose metabolism in humans is catalysed by CYP2C, in the rat are metabolised by the CYP3A subfamily; clearly, as far as the CYP2C subfamily is concerned the rat is not a suitable surrogate for man.

The CYP2C isoforms are modestly induced by drugs like phenobarbitone and dexamethasone in the liver of animals such as rat and rabbit, and experimental evidence has been presented that induction of this subfamily occurs also in humans exposed to rifampicin (Zhou *et al.* 1990). Chronic alcohol intake by rats has been shown to enhance the expression of CYP2C7, responsible for the 4-hydroxylation of retinoids, in the liver and colon (Hakkak *et al.* 1996). In rats, CYP2C11 was downregulated following administration of choline-deficient diets or vitamin A-deficient diets; supplementation of the animal diets with vitamin A and vitamin E resulted in a decrease in hepatic CYP2C6 activity and an increase in CYP2C11 activity (Ioannides 1998). CYP2C11 hepatic levels were reduced in the liver of rats following starvation or calorie restriction; starvation depressed also the levels of CYP2C13.

The human CYP2C subfamily appears to play no role in the bioactivation of chemical carcinogens (Shimada *et al.* 1989). In rats, however, members of the CYP2C family are efficient in converting aflatoxin B_1 to the reactive 8,9-epoxide (Shimada *et al.* 1987). The human CYP2C subfamily is responsible for the metabolic conversion of the diuretic drug tienilic acid to yield a protein-interacting intermediate, believed to be a sulphoxide, which can function as a hapten eliciting immunotoxicity (Pirmohamed and Park 1996).

The CYP2D subfamily

The realisation that the only human isoform identified, CYP2D6, is polymorphically expressed prompted further work in identifying the substrates of this enzyme, and led to studies of the pharmacogenetics of the human cytochrome P-450 proteins (Meyer *et al.* 1990). About 5–10% of Caucasians lack a functioning enzyme and are thus unable to metabolise the antihypertensive drug debrisoquine and other drugs that rely on this isoform for their metabolism. Consequently, such persons display an exaggerated response even following the intake of normal therapeutic doses of such drugs, or experience adverse effects commensurate with overdose, for example the vasodilator perhexiline causes peripheral neuropathy in poor metabolisers (Shah *et al.* 1982).

In addition to debrisoquine, CYP2D6 is a principal catalyst of the metabolism of other currently used major drugs including bufurarol, propranolol, atenolol, metoprolol, phenformin, nortriptyline, dextromethorphan and sparteine (Gonzalez 1996). In rat, CYP2D1 appears to be the responsible protein for the metabolism of these drugs.

Its role in chemical carcinogenesis is minimal; it has been shown to participate in the activation of the tobacco-specific nitrosamine NNK, but is not the principal catalyst (see above).

CYP2D proteins appear to be strictly constitutive and no inducer has so far been identified.

The CYP2E subfamily

In the rat and human this subfamily comprises a single isoform, CYP2E1, and the orthologues share the same substrate specificity. Indeed, the CYP2E subfamily, in common with the CYP1A, is one of the most conserved subfamilies in animal species. It is very active in the metabolism of small molecular weight compounds such as short-chain alcohols and organic solvents like ether and chloroform. CYP2E1 substrates are characterised by a small molecular diameter of <6.5 Å (Lewis *et al.* 1993). Its contribution to the metabolism of drugs is modest and is restricted to anaesthetics such as enflurane and other drugs like isoniazid and chlorzoxazone. A possible endogenous role is in the metabolism and elimination of acetone (Koop and Casazza, 1985), and the increase in activity seen in conditions of hyperketonaemia, for example insulin-dependent diabetes and starvation, may be viewed as an an adaptive response of the body to eliminate ketone bodies (Ioannides *et al.* 1996).

CYP2E plays a dominant role in the metabolism of small carcinogenic compounds such as azoxymethane, benzene, 1,3-butadiene, nitrosamines like dimethylnitrosamine and nitrosopyrrolidine, and halogenated hydrocarbons such as carbon tetrachloride and vinyl chloride (Guengerich *et al.* 1991). Moreover, it converts paracetamol (acetaminophen) to the hepatotoxic quinoneimine, and the anaesthetic halothane to yield the trifluoroacetyl intermediate which binds covalently to hepatic proteins, generating neoantigens that are responsible for the hepatitis associated with this drug (Pirmohamed and Park 1996). A major and toxicologically important characteristic of the CYP2E subfamily is its high propensity to generate reactive oxygen species (Ronis *et al.* 1996). In the rat, CYP2E1 was the most active in comparison with other isoforms in producing hydrogen peroxide (Albano *et al.* 1991). Reactive oxygen species are not only genotoxic, but can also stimulate cellular proliferation and dysplastic growth, thus accelerating the promotional stage of carcinogenesis. CYP2E may therefore facilitate carcinogenesis by two distinct mechanisms, namely the oxidative activation of chemicals and by the generation of reactive oxygen species.

CYP2E1 is an inducible protein, its levels being elevated following exposure to small molecular weight xenobiotics such as acetone, alcohol, pyrazole, imidazole, isoniazid, and in humans it is elevated in chronic alcoholics and following isoniazid intake (Tsutsume *et al.* 1989). In animal studies, CYP2E1 levels in the liver were upregulated when the diets were supplemented with corn oil, lard or medium chain triglycerides were administered intragastrically (Ioannides 1996). Thiamine deficiency also upregulates this isoform, whereas vitamin C deficiency leads to its downregulation. Starvation, even short term for only eight hours, as well as caloric restriction elevates the expression of this enzyme in the liver. A number of dietary anutrients have been shown to inhibit the activity of CYP2E1 in the liver including capsaicin and a number of organosulphates encountered in garlic (Ioannides 1999).

The CYP2F subfamily

This subfamily which appears to contain a single gene is poorly expressed in the human liver, but is present in the lung. It metabolises the lung toxin 3-methylindole (skatole) (Thornton-Manning *et al.* 1991). The role of this subfamily in drug metabolism and xenobiotic bioactivation remains to be clarified.

The CYP2G subfamily

CYP2G1 is a protein expressed only in the olfactory mucosa of animals. The murine orthologous protein has been shown recently to catalyse the bioactivation of the drug paracetamol (acetaminophen) and of the herbicide 2,6-dichlorobenzonitrile (Genter *et al.* 1998; Gu *et al.* 1998).

The CYP3 family

CYP3 is the most abundant family in human liver, the major protein being CYP3A4, and not surprisingly metabolism and deactivation of drugs is its principal function where it is particularly adept in catalysing *N*-dealkylation reactions (Rendic and Di Carlo 1997; Thummel and Wilkinson 1998). It is also present at high concentrations in the intestine (Kolars *et al.* 1992; McKinnon *et al.* 1995) where it is responsible for the first-pass metabolism of many drugs. Its substrate specificity is very broad, its substrate-binding site being sufficiently large to accommodate high molecular weight compounds like the macrolide antibiotics, but also versatile to contribute to the metabolism of small-size molecules such as paracetamol (acetaminophen). Not surprisingly, it is involved in many clinically relevant drug interactions (Thummel and Wilkinson, 1998). Two other genes are found in the liver of humans, CYP3A5 and A7. The latter is expressed at very high levels in the foetus, where it constitutes the major cytochrome P-450 enzyme, comprising some 50% of the cytochrome P-450 content in the liver, and it disappears shortly before birth and appears to be replaced by CYP3A4 (Schuetz *et al.* 1994; Lacroix *et al.* 1997).

Drug substrates comprise many major drugs including erythromycin, cyclosporin, dapsone, warfarin, tamoxifen, imipramine, prednisone, ketoconazole, miconazole, lidocaine, ethylmorphine, codeine, digitoxin, nifedipine, felodipine, warfarin and lovastatin as well as a number of steroid drugs (Maurel 1996). It also metabolises regioselectively steroid hormones such as ethinyl oestradiol to biologically inactive products, so that women taking CYP3A-inducing agents on a chronic basis, for example rifampicin to treat tuberculosis, conceived even when taking the contraceptive pill (Bolt 1994).

It is an inducible family both in rats and humans, and in the latter drugs like barbiturates, dexamethasone, erythromycin, phenobarbitone, rifampicin and troleandomycin were shown to elevate its expression (Maurel 1996). Levels of CYP3A2 in the liver were reduced in animals made cirrhotic by the administration of choline-deficient rats, but were increased following starvation (Ioannides 1998). A number of dietary anutrients increase the expression of CYP3A proteins such as indole-3-carbinol, organosulphates such as diallyl sulphide, but are decreased by phenethyl isothiocyanate (Ioannides 1999).

A number of carcinogenic chemicals have been shown to be bioactivated by the CYP3A subfamily, the most important being aflatoxin B_1 and dihydrodiols of polycyclic aromatic hydrocarbons in whose metabolism other cytochrome P-450 proteins also participate (Shimada et al. 1989; Kitada et al. 1990; Aoyama et al. 1990b). Members of the CYP3A family have also been shown to contribute to the bioactivation of the heterocyclic amine 2-amino-3-methylimidazo-[4,5-f]quinoline (IQ) (Kitada et al. 1990). The human CYP3A subfamily also catalyses the N-demethylation of cocaine leading to the generation of hepatotoxic metabolites (LeDuc et al 1993; Pasanen et al. 1995).

The CYP4 family

Three subfamilies have been identified in humans and rats, CYP4A, 4B and 4F, but the catalytic activity of these towards xenobiotics has not been clearly defined. No major role in xenobiotic metabolism has so far been ascribed to this family, its principal function being in the metabolism of endobiotics such as fatty acids and eicosanoids. An interesting aspect of this family is its induction by peroxisomal proliferators, epigenetic carcinogens that increase the number and size of peroxisomes and cause cellular proliferation. It has now been established that the induction of the CYP4A subfamily and peroxisomal proliferation are coordinately regulated by the same receptor (Johnson et al. 1996).

In the rabbit, a protein expressed principally in the lung, CYP4B1, activates a number of aromatic amines, such as 2-aminofluorene, through N-hydroxylation, and the pulmonary toxin ipomeanol (Czerwinski et al. 1991; Verschoyle et al. 1993). The human orthologue, however, is unable to

Table 4.4. Characteristics of the xenobiotic-metabolising cytochrome P450 enzymes

Cytochrome P-450 subfamily	Role in drug metabolism	Examples of drug substrates	Role in chemical carcinogenesis	Examples of carcinogens activated	Inducibility	Other characteristics
CYP1A	Moderate	Theophylline Imipramine	Very extensive	Benzo[a]pyrene 4-Aminobiphenyl	Very high	Substrates essentially planar Regulated by the Ah receptor
CYP1B	Not evaluated	?	Very extensive	2-Aminoanthracene 2-Nitropyrene	Very high	Substrates essentially planar Regulated by the Ah receptor
CYP2A	Limited	Coumarin Propranolol	Moderate	6-Aminochrysene 2-Nitropyrene	Moderate	Human and rat proteins differ in substrate specificity
CYP2B	Limited	Benzphetamine Cocaine	Moderate	Cyclophosphamide NNK[a]	High	Catalyses the deactivation of many carcinogens
CYP2C	Extensive	Tolbutamide Mephenytoin	None	—	Poor	Polymorphically expressed
CYP2D	Extensive	Debrisoquine Dextromethorphan	Poor	NNK	Not inducible	Polymorphically expressed
CYP2E	Limited	Enflurane Paracetamol	Extensive	Carbon tetrachloride Dimethylnitrosamine	High	Propensity to generate reactive oxygen species
CYP2F	Not evaluated	?	Not evaluated	?	?	Principally expressed in the lung
CYP2G	Not evaluated	Paracetamol	Not evaluated	?	?	Principally expressed in the olfactory mucosa
CYP3A	Very extensive	Dapsone Cyclosporin Midazolam	Moderate	Aflatoxin B₁ Senecionine	High	Most abundant subfamily in the human liver
CYP4A	Minimal	Midazolam	None	—	High	Its induction is linked to peroxisomal proliferation

[a]NNK, 4-(Methylnitrosamino)-1-(3-pyridyl)-1-butanone.

carry out these reactions. It also contributes to the metabolism of midazolam (Wandel *et al.* 1994).

In the liver, CYP4A activity is enhanced by fibrate hypolipidaemic drugs such as clofibrate and ciprofibrate, and anti-inflammatories such as benoxaprofen and ibuprofen (Lake and Lewis 1996; Rekka *et al.* 1994). The naturally occurring terpenoid, citral, as well as alcohol stimulate hepatic CYP4A1 levels; similarly higher levels have been reported in animals treated with medium chain triglycerides (Ioannides 1999). CYP4B has been shown to be inducible by phenobarbital, at least in the rabbit (Robertson *et al.* 1983).

The major characteristics of the xenobiotic-metabolising cytochromes P-450 are summarised in Table 4.4.

Cytochrome P-450 induction and chemical carcinogenesis: the Ah receptor

Most xenobiotic-metabolising cytochrome P-450 isoforms can be induced by previous exposure to a variety of structurally unrelated xenobiotics including many drugs and environmental contaminants. Diet is a rich source of inducing agents (Ioannides 1999), and consumption of cruciferous vegetables or charcoal-broiled beef stimulated the metabolism of drugs such as phenacetin and caffeine (Conney *et al.* 1977; Kall *et al.* 1996). Certain isoforms, however, such as the members of the CYP2D subfamily appear to be strictly constitutive, with no inducers being identified so far.

Induction of the cytochrome P-450 enzymes can occur in all tissues, but the liver appears to be the most sensitive. Induction of extrahepatic cytochromes P-450 may be governed by the route of administration of the inducing agent; gastrointestinal cytochromes P-450 are more likely to be induced when the inducer is administered orally, whereas the pulmonary proteins will be upregulated by inducers taken through inhalation. The extent of induction, at least in the case of barbiturates, appears to be closely related to the plasma half-life of the drug; phenobarbitone, an inducer of CYP2B and 3A enzymes, having a long half-life is one of the most potent inducers of cytochromes P-450, whereas barbiturates with a short half-life such as hexobarbitone are very poor inducing agents (Ioannides and Parke 1974). Whether the same relationship holds for other cytochrome P-450 isoforms, remains to be established.

Of the various cytochrome P-450 families, CYP1 is the most inducible (see above), its activity being increased many-fold in the liver and other tissues following administration of potent inducing agents; in the liver of rats, the CYP1 family after induction may comprise as much as 80% of the total cytochrome P-450 content (Ryan and Levin 1990). Bearing in mind that CYP1 is the predominant family in the activation of carcinogenic chemicals to which humans are continuously and involuntarily exposed to, what is the effect of induction on chemical carcinogenesis?

The CYP1 is a minor family in the liver of humans and animals; as a result the rate of production of reactive intermediates is slow, and the levels of reactive intermediates generated low, so that these are readily detoxicated by defensive systems such as glutathione conjugation. Under such circumstances no cellular damage will ensue. Indeed, in *in vitro* mutagenicity assays such as the Ames test, rat and human preparations, when employed as activation systems, fail to provoke a mutagenic response with many established human chemical carcinogens such as benzo(a)-pyrene, 4-aminobiphenyl and 2-naphthylamine. It is for this reason that for such tests to be able to identify these carcinogens, it is essential to utilise as activation system hepatic preparations from animals pretreated with inducing agents, the most common being Aroclor 1254, to boost cytochrome P-450 activity, and especially CYP1A. The high CYP1A activity ensures that activation pathways assume greater importance and the rate of generation of reactive intermediates is markedly elevated.

Since the CYP1A levels are so low, how are chemicals converted to the reactive intermediates that manifest their carcinogenicity? It is conceivable that the carcinogens themselves can, on repeated administration, stimulate selectively their own activation pathways, leading to higher generation of reactive intermediates which, having overwhelmed the detoxication processes, are available to induce DNA damage and initiate the carcinogenic process. The detrimental role of elevated CYP1 expression in chemical carcinogenesis is supported by extensive experimental studies (Nebert 1989, 1991; Ioannides and Parke 1993). For example, DBA/2 mice, a strain refractive to CYP1A induction, was less susceptible to the carcinogenicity of the CYP1A1 substrate benzo[a]pyrene compared with the C57/BL strain, where CYP1A activity is more readily inducible (Nebert 1989). In another animal study, of 10 generations of rats exposed to the CYP1A-activated carcinogen 3'-methyl-4-dimethylaminoazobenzene, the fourth to eighth generations were resistant to the carcinogenicity of this chemical compared to the other generations; the resistant generations also displayed poor CYP1A inducibility and, consequently, their ability to bioactivate this carcinogen was low (Yoshimoto *et al.* 1985; Yano *et al.* 1989). Sex variations in carcinogenic response have also been ascribed to differences in CYP1 inducibility; female rodents were more sensitive to the carcinogenicity of some aminoacid pyrolysis products and, moreover, exhibited a higher degree of CYP1 inducibility (Degawa *et al.* 1987).

Induction of cytochrome P-450 activity leading to enhanced toxicity and carcinogenicity is not limited to the CYP1A family, but has also been established with inducers of the CYP2E1 isoform. Humans exposed to alcohol, the prototype inducer of CYP2E1, on a chronic basis were more susceptible to the toxicity of paracetamol (acetaminophen) a substrate of this isoform (Zimmerman and Maddrey 1995). In animal models, prolonged intake of alcohol has been shown to exacerbate the toxicity and carcinogenicity of many CYP2E substrates including carbon tetrachloride (Lieber 1991).

Collectively, the above observations raise the possibility that the ability of a chemical to stimulate its own activation pathways may be a critical factor in determining its carcinogenic activity (Ioannides 1990). Preliminary studies indicated that, among isomers, carcinogenic potency could be related to ability to induce CYP1A activity. For example benzo[a]pyrene, 4-aminobiphenyl and 2-naphthylamine were far more potent inducers of hepatic CYP1A than their non-carcinogenic isomers benzo[e]pyrene, 2-aminobiphenyl and 1-naphthylamine (Ayrton et al. 1990a, b). In order to further define this relationship, extensive studies were undertaken to correlate CYP1A induction, binding to the Ah receptor, the cytosolic receptor regulating CYP1 expression (see above) and carcinogenic activity among isomers or structurally related compounds. Such studies revealed that treatment of animals with the carcinogenic 4-acetylaminofluorene resulted in the induction of both CYP1A isoforms, whereas the 2-isomer increased only the levels of CYP1A2; in concordance with these findings, the 4-isomer bound to the Ah receptor markedly more avidly compared to 2-acetylaminofluorene (Ioannides et al. 1993). The consequence of the increased CYP1A expression was that 2-acetylaminofluorene stimulated markedly its own activation to mutagenic metabolites, whereas the 4-isomer had only a modest effect. Moreover, in extensive subsequent studies involving six azobenzenes, a correlation was established between binding to the Ah receptor and CYP1A induction on the one hand, and carcinogenic activity on the other. In this case, however, induction of CYP1A did not translate into increased production of mutagenic intermediates (Cheung et al. 1994). A similar picture emerged when isomeric diaminotoluenes or diaminonaphthalenes and related compounds were employed as the model compounds (Cheung et al. 1996, 1997). Indeed, the first observations that CYP1A induction by CYP1A-activated aminocompounds does not always lead to increase in the rate of activation were first made in 1986 (Steele and Ioannides 1986) using aromatic amines as test compounds. Collectively, the above experimental findings led to the conclusion that the critical event in the carcinogenicity of planar compounds is the binding to the Ah receptor and not the ensuing CYP1 induction, since the latter may not be manifested as enhanced activation. In addition to CYP1A induction, the Ah receptor mediates many other activities including protein kinase C, whose activation culminates in accelerated DNA replication, dedifferentiation and cellular proliferation, a critical aspect of the carcinogenesis process (Droms and Malkinson 1991; Sogawa and Fujii-Kuriyama 1997). It may be, therefore, that it is the increase in promotion and progression, mediated by the Ah receptor, which is the most relevant to chemical carcinogenesis, and that CYP1A induction is fortuitous and of secondary importance in determining carcinogenic potential. Interaction with the Ah receptor may be thus utilised as an early screen for carcinogenic potential.

Conclusions

The initial view that the cytochrome P-450 enzyme system functions simply in the metabolism of xenobiotics to readily excretable metabolites to enable the elimination of these from the body is anachronistic on the face of mounting evidence that this system can also convert innocuous chemicals to deleterious products, detrimental to the welfare of the living organism. There are numerous examples documented where cytochrome P-450 metabolism resulted in toxicity. However, not all xenobiotic-metabolising P-450 proteins participate in the bioactivation of chemicals; there is a marked difference in the role of individual cytochromes P-450 in xenobiotic activation. For example, the CYP2C, 2B and 2D subfamilies play virtually no role in the bioactivation of toxic and carcinogenic chemicals and appear to metabolise these to inactive products, whereas the CYP1A, 1B and 2E subfamilies are responsible for the bioactivation of the vast majority of xenobiotics. Consequently, the fate of a chemical in the body, whether activation or deactivation, is dependent on factors such as the following:

1. The concentration and profile of cytochromes P-450.
2. The tissue concentration of the chemical. At high tissue concentrations, more than one form of cytochrome P-450 may be involved, as low-affinity forms may also start contributing significantly to metabolism.
3. The levels of enzymes that detoxicate reactive intermediates, for example glutathione S-transferases, and reactive oxygen species, such as superoxide dismutase. Any factor that modulates the enzymes involved in the metabolism of a certain chemical will also influence its toxicity.

It has been frequently pointed out that too many chemicals come out as positive in long-term carcinogenicity studies in rodents, where excessive doses, equivalent to the maximum tolerated dose, are employed (Ames and Gold 1998). At high doses tissue damage and inflammation may occur triggering cell division, a critical event in the carcinogenesis process (Gold et al. 1998). An additional mechanism that may contribute in the high rate of positives is cytochrome P-450 induction. In carcinogenicity studies, the evaluated chemical is administered, not only at high doses, but also daily. Such a treatment, after a few doses of the chemical, may lead to a selective induction of the bioactivating enzymes, thereby exaggerating the production of reactive intermediates, overwhelming the detoxication processes, and markedly increasing the likelihood of a positive carcinogenic effect. It is unlikely that lower doses and/or intermittent intake that occurs in humans will result in cytochrome P-450 induction.

Clearly, induction of cytochromes P-450 is an important phenomenon that may predispose humans to the carcinogenicity of chemicals. Although cytochrome P-450 induction is seen readily in animals, it is much less common in humans, and this may be one of the reasons why humans survive in an environment littered with toxic and carcinogenic pollutants.

Acknowledgement

The author is grateful to Miss Sheila Evans for drawing all the figures.

References

Ahmad, N., Katiyar, S.K. and Mukhtar, H. (1998) Cancer chemoprevention by tea polyphenols. In: *Nutrition and Chemical Toxicity*, C. Ioannides, ed. John Wiley, Chichester, pp. 301–343.

Albano, E., Tomasi, A., Persson, J.-Q., Terelius, Y., Goria-Gatti, I., Ingelman-Sundberg, M. and Dianzani, M.U. (1991) Role of ethanol-inducible cytochrome P450 (P450IIE1) in catalysis of the free radical activation of aliphatic alcohols. *Biochem. Pharmacol.*, **41**, 1895–1902.

Ames, B.N. and Gold, L.S. (1998) The prevention of cancer. *Drug Metab. Rev.*, **30**, 201–223.

Anttila, S., Hietanen, E., Vainio, H., Camus, A.-M., Gelboin, H.V., Park, S.S., Heikkila, I., Karjalainen, A. and Bartsch, H. (1991) Smoking and peripheral type of cancer are related to high levels of pulmonary cytochrome P450IA in lung cancer patients. *Int. J. Cancer*, **47**, 681–685.

Anttila, S., Vainio, H., Hietanen, H., Camus, A-M., Malaveille, C., Brun, G., Husgafvel-Pursiainen, K., Heikkila, L., Karjalainen, A. and Bartsch, H. (1992) Immunohistochemical detection of pulmonary cytochrome P450IA and metabolic activities associated with P450IA1 and P450IA2 isozymes in lung cancer patients. *Environ. Health Perspect.*, **98**, 179–182.

Aoyama, T., Korzekwa, K., Nagata, K., Gillette, J., Gelboin, H.V. and Gonzalez, F.J. (1990a) Estradiol metabolism by complementary deoxyribonucleic acid-expressed human cytochrome P450s. *Endocrinology*, **126**, 3101–3106.

Aoyama, T., Yamano, S., Guzelian, P.S., Gelboin, H.V. and Gonzalez, H.V. (1990b) Five of 12 forms of vaccinia virus-expressed human hepatic cytochrome P450 metabolically activate aflatoxin B$_1$. *Proc. Nat. Acad. Sci. US*, **87**, 4790–4793.

Astrom, A. and De Pierre, J.W. (1985) Metabolism of 2-acetylaminofluorene by eight different forms of cytochrome P-450 isolated from rat liver. *Carcinogenesis*, **6**, 113–120.

Ayrton, A. D., McFarlane, M., Walker, R., Neville, S. and Ioannides, C. (1990a) The induction of P450 I proteins by aromatic amines may be related to their carcinogenic potential. *Carcinogenesis*, **11**, 803–809.

Ayrton, A.D., McFarlane, M., Walker, R., Neville, S., Coombs, M.M. and Ioannides, C. (1990a) Induction of the P450 I family of proteins by polycyclic aromatic hydrocarbons. Possible relationship to their carcinogenicity. *Toxicology*, **60**, 173–186.

Bailey, D.G., Malcolm, J., Arnold, O. and Spence, J.D. (1998) Grapefruit juice–drug interactions. *Br. J. Clin. Pharmacol.*, **46**, 101–110.

Bartsch, H., Castegnaro, M., Rojas, M., Camus, A.-M., Alexandrov, K. and Lang, M. (1992) Expression of pulmonary cytochrome P4501A1 and carcinogen adduct formation in high risk subjects for tobacco-related lung cancer. *Toxicol. Lett.*, **64/65**, 477–483.

Beaune, P.H. and Bourdi, M. (1993) Autoantibodies against cytochrome P-450 in drug-induced autoimmune hepatitis. *Ann. New York Acad. Sci.*, **658**, 641–645.

Bolt, H.M. (1994) Interactions between clinically used drugs and oral contraceptives. *Environ. Health Perspect.*, **102**, 35–38.

Chang, T.K.H. and Waxman, D.J. (1996) The CYP2A subfamily. In: *Cytochromes P450: Metabolic and Toxicological Aspects*, C. Ioannides, ed. CRC Press, Boca Raton, FL, pp. 99–134.

Chang, T.K.H., Weber, G.F., Crespi, C.L. and Waxman, D.J. (1993) Differential activation of cyclophosphamide and ifosphamide by cytochromes P450 2B and 3A in human liver microsomes. *Cancer Res.*, **53**, 5629–5637.

Cheung, Y.-L., Lewis, D.F.V., Ridd, I., Gray, T.J.B. and Ioannides, C. (1997) Diaminonaphthalenes and related compounds: mutagenicity, CYP1A induction and interaction with the Ah receptor. *Toxicology*, **118**, 115–127.

Cheung, Y.-L., Puddicombe, S.M., Gray, T.J.B. and Ioannides, C. (1994) Mutagenicity and CYP1A induction by azobenzenes correlates with their carcinogenicity. *Carcinogenesis*, **15**, 1257–1263.

Cheung, Y.-L., Snelling, J., Mohammed, N.N.D., Gray, T.J.B. and Ioannides, C. (1996) Interaction with the aromatic hydrocarbon receptor, CYP1A induction, and mutagenicity of a series of diaminotoluenes: implications for their carcinogenicity. *Toxicol. and Appl. Pharmacol.*, **139**, 203–211.

Code, E.L., Crespi, C.L., Penman, B.W., Gonzalez, F.J., Chang, T.K.H. and Waxman, D.J. (1997) Human cytochrome P4502B6. Interindividual hepatic expression, substrate specificity, and role in procarcinogen activation. *Drug Metab. and Disposition*, **25**, 985–993.

Conney, A.H., Pantuck, E.J., Kuntzman, R., Kappas, A., Anderson, K.E. and Alvares, A.P. (1977) Nutrition and chemical biotransformation in man. *Clin. Pharmacol. and Ther.*, **22**, 707–719.

Crespi, C.L., Penman, B.W., Gelboin, H.V. and Gonzalez, F.J. (1991) A tobacco smoke-derived nitrosamine, 4-(methylnitrosamino)-*l*-(3-pyridyl)-*l*-butanone, is activated by multiple human cytochromes P450 including the polymorphic human cytochrome CYP2D6. *Carcinogenesis*, **12**, 1197–1201.

Crespi, C.L., Penman, B., Leake, J.A.E., Arlotto, M., Stark, A., Turner, T., Steimel, D., Rudo, K., Davies, R.L. and Langenbach, R. (1990) Human cytochrome P450IIA3 cDNA sequence, role of the enzyme in the metabolic activation of promutagens, comparison to nitrosamine activation by human cytochrome P450IIE1. *Carcinogenesis*, **11**, 1293–1300.

Czerwinsky, M., McClemore, T.L., Philpot, R.M., Nhamburo, P.T., Korzekwa, K., Gelboin, H.V. and Gonzalez, F.J. (1991) Metabolic activation of 4-ipomeanol by complementary DNA-expressed human cytochromes P-450: evidence for species-specific metabolism. *Cancer Res.*, **51**, 4636–4648.

Davis, C.D., Ghoshal, A., Schut, H.A.J. and Snyderwine, E.G. (1994) Metabolic activation of heterocyclic amine food mutagens in the mammary gland of lactating Fischer 344 rats. *Cancer Lett.*, **84**, 67–73.

De Matteis, F., Dawson, S.J., Boobis, A.R. and Comoglio, A. (1991) Inducible bilirubin-degrading system of rat liver microsomes: role of cytochrome P450IA1. *Molec. Pharmacol.*, **40**, 686–691.

De Waziers, I., Cugnenc, P.H., Yang, C.S., Leroux, J.-P. and Beaune, P.H. (1990) Cytochrome P450 isoenzymes, epoxide hydrolase in rat and human hepatic and extrahepatic tissues. *J. Pharmacol. and Exp. Ther.*, **253**, 387–394.

Degawa, M., Hishimuma, T., Koshida, H. and Hashimoto, Y. (1987) Species, sex and organ differences in induction of a cytochrome P-450 isozyme responsible for carcinogen activation: effects of dietary hepatocarcinogenic tryptophan pyrolysate components in mice and rats. *Carcinogenesis*, **8**, 1913–1918.

Diaz, D., Fabre, I., Daujat, M., Saint Aubert, B., Michel, H. and Maurel, P. (1990) Omeprazole is an aryl hydrocarbon-like inducer of hepatic cytochrome P450. *Gastroenterology*, **99**, 737–747.

Droms, K.A. and Malkinson, A.M. (1991) Phorbol ester-induced tumor promotion by downregulation of protein kinase C. *Molec. Carcinog.*, **4**, 1–2.

Duescher, R.J. and Elfarra, A.A. (1994) Human liver microsomes are efficient catalysts of 1,3-butadiene oxidation: evidence of major roles by cytochrome P4502A6 and 2E1. *Arch. Biochem. and Biophys.*, **311**, 342–349.

Eaton, D.L., Gallagher, E.P., Bammler, T.K. and Kunze, K.L. (1995) Role of cytochrome P4501A2 in chemical carcinogenesis: implications for human variability in expression and enzyme activity. *Pharmacogenetics*, **5**, 259–274.

Estabrook, R.W. (1996) Cytochrome P450: from a single protein to a family of proteins – with some personal reflections. In: *Cytochromes P450: Metabolic and Toxicological Aspects*, C. Ioannides, ed. CRC Press, Boca Raton, FL, pp. 3–29.

Genter, M.B., Liang, H-C., Gu, J., Ding, X., Negishi, M., McKinnon, R.A. and Nebert, D.W. (1998) Role of CYP2A5 and 2G1 in acetaminophen metabolism and toxicity in the olfactory mucosa of the *Cyp1a2* $(-/-)$ mouse. *Biochem. Pharmacol.*, **55**, 1819–1826.

Gold, L.S., Slone, T.H. and Ames, B.N. (1998) What do animal cancer tests tell us about human cancer risk?: Overview of analyses of the carcinogenic potency database. *Drug Metab. Rev.*, **30**, 359–404.

Goldstein, J.A. and de Morais, S.M.F. (1994) Biochemistry and molecular biology of the human CYP2C subfamily. *Pharmacogenetics*, **4**, 285–299.

Gonzalez, F.J. (1996) The CYP2D subfamily. In *Cytochromes P450: Metabolic and Toxicological Aspects*, C. Ioannides, ed. CRC Press, Boca Raton, FL, pp. 183–210.

Gonzalez, F.J. and Gelboin, H.V. (1994) Human cytochromes P450: evolution and cDNA-directed expression. *Environ. Health Perspect.*, **98**, 81–85.

Gonzalez, F.J. and Nebert, D.W. (1990) Evolution of the P450 gene superfamily animal plant 'warfare', molecular drive and human differences in drug oxidation. *Trends in Genetics*, **6**, 182–186.

Gu, J., Zhang, Q.-Y., Genter, M.B., Lipinskas, T.W., Negishi, M., Nebert, D.W. and Ding, X. (1998) Purification and characterization of heterologously expressed mouse CYP2A5 and CYP2G1: role in metabolic activation of acetaminophen and 2,6-dichlorobenzonitrile in mouse olfactory mucosal microsomes. *J. Pharmacol. and Exp. Ther.*, **285**, 1287–1295.

Guengerich, F.P. (1990) Characterization of roles in human cytochrome P-450 enzymes in carcinogen metabolism. *Asia Pacific J. Pharmacol.*, **5**, 327–345.

Guengerich, F.P. (1993) Cytochrome P450 enzymes. *Am. Sci.*, **81**, 440–447.

Guengerich, F.P. and Shimada, T. (1991) Oxidation of toxic and carcinogenic chemicals by human cytochrome P-450 enzymes. *Chem. Res. Toxicol.*, **4**, 391–407.

Guengerich, F.P., Gilliam, E.M.J. and Shimada, T. (1996) New applications of bacterial systems to problems in toxicology. *Crit Rev. Toxicol.*, **26**, 551–583.

Guengerich, F.P., Kim, D.H. and Iwasaki, M. (1991) Role of human cytochrome P-450 IIE1 in the oxidation of many low molecular weight cancer suspects. *Chem. Res. Toxicol.*, **4**, 168–179.

Hakkak, R., Korounian, S., Ronis, M.J., Ingelman-Sundberg, M. and Badger, T.M. (1996) Effects of diet and ethanol on the expression and localization of cytochromes P450 2E1 and P450 2C7 in the colon of male rats. *Biochem. Pharmacol.*, **51**, 61–69.

Hammons, G.J., Guengerich, F.P., Weis, C.C., Beland, F.A. and Kadlubar, F.F. (1985) Metabolic oxidation of carcinogenic arylamines by rat, dog, and human hepatic microsomes and by purified flavin-containing and cytochrome P-450 mono-oxygenases. *Cancer Res.*, **45**, 3578–3585.

Hankinson, O. (1995) The aryl hydrocarbon receptor complex. *Ann. Rev. Pharmacol. and Toxicol.*, **35**, 307–340.

Hayes, C.L., Spink, D.C., Spink, B.C., Cao, J.Q., Walker, N.J. and Sutter, T.R. (1996) 17β-Estradiol hydroxylation catalysed by human cytochrome P4501B1. *Proc. Nat. Acad. Sci. US*, **93**, 9776–9781.

Honkakoski, P. and Negishi, M. (1997) The structure, function, and regulation of cytochrome P450 2A enzymes. *Drug Metab. Rev.*, 29, 977–996.

Imaoka, S., Terano, Y. and Funae, Y. (1990) Changes in the amount of cytochromes P450 in rat hepatic microsomes with starvation. *Arch. of Biochem. and Biophys.*, **278**, 168–178.

Ingelman-Sundberg, M. and Johansson, I. (1984) Mechanisms of hydroxyl radical formation and ethanol oxidation by ethanol-inducible and other forms of rabbit liver microsomal cytochromes P-450. *J. Biol. Chem.*, **259**, 6447–6458.

Ioannides, C. (1990) Induction of cytochrome P450 I and its influences in chemical carcinogenesis. *Biochem. Soc. Trans.*, **18**, 32–34.

Ioannides, C. (1998) Nutritional modulation of cytochromes P450. In *Nutrition and Chemical Toxicity*, C. Ioannides, ed. John Wiley, Chichester, pp. 115–159.

Ioannides, C. (1999) Effect of diet and nutrition on the expression of cytochromes P450. *Xenobiotica*, **29**, 109–154.

Ioannides, C. and Parke, D.V. (1975) Mechanism of induction of hepatic drug metabolising enzymes by a series of barbiturates. *J. Pharm. and Pharmacol.*, **27**, 739–749.

Ioannides, C. and Parke, D.V. (1990) The cytochrome P450 I gene family of microsomal haemoproteins and their role in the metabolic activation of chemicals. *Drug Metabol. Rev.*, **22**, 1–85.

Ioannides, C. and Parke, D.V. (1993) Induction of cytochrome P4501 as an indicator of potential chemical carcinogenesis. *Drug Metabol. Rev.*, **25**, 485–501.

Ioannides, C., Delaforge, M. and Parke, D.V. (1981) Safrole: its metabolism, carcinogenicity and interactions with cytochrome P-450. *Food and Cosmetics Toxicology*, **19**, 657–666.

Ioannides, C., Barnett, C.R. Irizar, A. and Flatt, P.R. (1996) Expression of cytochrome P450 proteins in disease. In: *Cytochromes P450: Metabolic and Toxicological Aspects*, C. Ioannides, ed. CRC Press, Boca Raton, FL, pp. 301–327.

Ioannides, C., Cheung, Y.-L., Wilson, J., Lewis, D.F.V. and Gray, T.J.B. (1993) The mutagenicity and interactions of 2- and 4-(acetylamino)fluorene with cytochrome P450 and the aromatic hydrocarbon receptor may explain the difference in their carcinogenic potency. *Chem. Res. in Toxicol.*, **6**, 535–541.

Ioannides, C., Lewis, D.F.V., Trinick, J., Neville, S., Sertkaya, N. N., Kajbaf, M. and Gorrod, J.W. (1989) A rationale for the non-mutagenicity of 2- and 3-amino-biphenyls. *Carcinogenesis*, **10**, 1403–1407.

Jewell, H., Maggs, J.L., Harrison, A.C., O'Neill, M.M., Ruscoe, J.E. and Park, B.K. (1995) Role of hepatic metabolism in the bioactivation and detoxication of amodiaquine. *Xenobiotica*, **25**, 199–217.

Jin, L. and Baillie, T.A. (1997) Metabolism of the chemopreventive agent diallyl sulphide to glutathione conjugates in rats. *Chem. Res. in Toxicol.*, **10**, 318–327.

Johnson, E.F., Palmer, C.N.A., Griffin, K.J. and Hse, M.-H. (1996) Role of the peroxisome proliferator-activated receptor in cytochrome P450 4A gene regulation. *FASEB J.*, **10**, 1241–1248.

Kaderlik, K.R., Minchin, R.F., Mulder, G.J., Ilett, K.F., Daugaard-Jenson, M., Teitel, C.H. and Kadlubar, F.F. (1994) Metabolic activation pathway for the formation of DNA adducts of the carcinogen 2-amino-1-methyl-6-phenylimidazo[4,5-b]pyridine (PhIP) in rat extrahepatic tissues. *Carcinogenesis*, **15**, 1703–1709.

Kadlubar, F.F., Unruh, L.E., Flammang, T.J., Sparks, D., Mitchum, R.K. and Mulder, G.J. (1981) Alteration of urinary levels of the carcinogen N-hydroxy-2-naphthylamine, and its N-glucuronide in the rat by control of urinary pH, inhibition of metabolic sulfation and changes in biliary excretion. *Chemico-Biol. Interactions*, **33**, 129–147.

Kall, M.A., Vang, O. and Clausen, J. (1996) Effects of dietary broccoli in human *in vivo* drug metabolizing enzymes: evaluation of caffeine, oestrone and chlorzoxazone metabolism. *Carcinogenesis*, **17**, 793–799.

Kawajiri, K. and Hayashi, S.-I. (1996) The CYP1 family. In: *Cytochromes P450: Metabolic and Toxicological Aspects*, C. Ioannides, ed. CRC Press, Boca Raton, FL, pp. 77–97.

Kawajiri, K., Yonekawa, H., Hara, E. and Tagashira, Y. (1978) Biochemical basis for the resistance of guinea pigs to carcinogenesis by 2-acetylaminofluorene. *Biochem. Biophys. Res. Commun.*, **85**, 959–965.

Kellermann, G., Shaw, C.R. and Luyten-Kellermann, M. (1973) Aryl-hydrocarbon hydroxylase inducibility and bronchogenic carcinoma. *New England J. Med.*, **289**, 934–937.

Kim, J.H., Stansbury, K.H., Walker, N.J., Trush, M.A., Strinkland, P.T. and Sutter, T.R. (1998) Metabolism of benzo[a]pyrene and benzo[a]pyrene-7,8-diol by human cytochrome P450 1B1. *Carcinogenesis*, **19**, 1847–1853.

Kimura, S., Kozak, C.A. and Gonzalez, F.J. (1989) Identification of a novel P450 expressed in rat lung: cDNA cloning and sequence, chromosome mapping, and induction by 3-methylcholanthrene. *Biochemistry*, **28**, 3798–3803.

Kitada, M., Taneda, M., Ohta, K., Nagashima, K., Itahashi, K. and Kamataki, T. (1990) Metabolic activation of aflatoxin B_1 and 2-amino-3-methylimidazo[4,5-f]quinoline by human adult and fetal livers. *Cancer Res.*, **50**, 2641–2645.

Knodell, R.G., D., Gwodz, G.P., Brian, W.R. and Guengerich, F.P. (1991) Differential inhibition of human liver cytochromes P-450 by cimetidine. *Gastroenterology*, **101**, 1680–1691.

Kolars, J.C., Schmiedlin-Ren, P., Schuetz, J.D., Fang, C. and Watkins, P.B. (1992) Identification of rifampin-inducible P450IIIA4 (CYP3A4) in human small bowel enterocytes. *J. of Clin. Investigation*, **90**, 1871–1878.

Koop, D.R. and Casazza, J.P. (1985) Identification of the ethanol-inducible P-450 isozyme 3a as the acetone and acetol monooxygenase of rabbit microsomes. *J. Biol. Chem.*, **260**, 13607–13612.

Korsgaard, R., Trell, E., Sinonsson, B.G. Stiksa, G., Janzon, L., Hood, B. and Oldbring, J. (1984) Aryl hydrocarbon hydroxylase induction levels in patients with malignant tumors associated with smoking. *J. Cancer Res. and Clin. Oncol.*, **108**, 286–289.

Kouri, R.E., Mckinney, C.E., Slomiany A.J., Snodgrass, D.R., Wray, N.P. and McLemore, T.L. (1982) Positive correlations between high aryl hydrocarbon hydroxylase and primary lung cancer as analyzed in cryo-preserved lymphocytes. *Cancer Res.*, **42**, 5030–5037.

Lake, B.G. and Lewis, D.F.V. (1996) The CYP4 family. In: *Cytochromes P450: Metabolic and Toxicological Aspects*, C. Ioannides, ed. CRC Press, Boca Raton, FL, pp. 271–297.

LeDuc, B.W., Sinclair, P.R., Shuster, L., Sinclair, J.F., Evans, J.E. and Greenblatt, D.J. (1992) Norcocaine and N-hydroxynorcocaine formation in human liver microsomes: role of cytochrome P-450 3A4. *Pharmacology*, **46**, 294–300.

Lewis, D.F.V., Ioannides, C. and Parke, D.V. (1986) Molecular dimensions of the substrate binding site of cytochrome P-448. *Biochem. Pharmacol.*, **35**, 2179–2185.

Lewis, D.F.V., Ioannides, C. and Parke, D.V. (1987) Structural requirements for substrates of cytochromes P-450 and P-448. *Chemico-Biol. Interactions*, **64**, 39–60.

Lewis, D.F.V., Ioannides, C. and Parke, D.V. (1993) Validation of a novel molecular orbital approach (COMPACT) for the prospective safety evaluation of chemicals by comparison with rodent carcinogenicity and Salmonella mutagenicity data evaluated by the US NCI/NTP. *Mutat. Res.*, **291**, 61–77.

Lewis, D.F.V., Ioannides, C. and Parke, D.V. (1994) Molecular modelling of cytochrome CYP1A1: a putative access channel explains differences in induction potency between the isomers benzo(α)pyrene and benzo(e)pyrene and 2- and 4-acetylaminofluorene. *Toxicol. Lett.*, **71**, 235–243.

Lewis, D.F.V., Watson, E. and Lake, B.G. (1998) Evolution of the cytochrome P450 superfamily: Sequence alignments and pharmacogenetics. *Mutat. Res.*, **410**, 245–270.

Lieber, C.S. (1991) Hepatic, metabolic and toxic effects of ethanol: 1991 update. *Alcoholism: Clin. and Exp. Res.*, **15**, 573–592.

Lindberg, R.J.P. and Negishi, M. (1989) Alteration of mouse cytochrome P450$_{coh}$ substrate specificity by mutation of a single amino-acid residue. *Nature*, **339**, 632–634.

Liu, C., Zhuo, X., Gonzalez, F.J. and Ding, X.X. (1996) Baculovirus-mediated expression and characterization of rat CYP2A3 and human CYP2A6: role in metabolic activation of nasal toxicants. *Mol. Pharmacol.*, **50**, 781–788.

Lum, P.Y., Walker, S. and Ioannides, S. (1985) Foetal and neonatal development of cytochrome P-450 and cytochrome P-448 catalysed mixed-function oxidases in the rat: induction by 3-methylcholanthrene. *Toxicology*, **35**, 307–317.

Maurel, P. (1996) The CYP3 family. In *Cytochromes P450: Metabolic and Toxicological Aspects*, C. Ioannides, ed. CRC Press, Boca Raton, FL, pp. 241–270.

McDonell, W.M., Scheiman, J.M. and Traber, P.G. (1992) Induction of cytochrome P4501A genes (*CYP1A*) by omeprazole in the human alimentary tract. *Gastroenterology*, **103**, 1509–1516.

McKinnon, R.A., Burgess, W.M., Hall, M.P., Roberts-Thomson, S.J., Gonzalez, F.J. and McManus, M.E. (1995) Characterization of CYP3A gene subfamily expression in human gastrointestinal tissues. *Gut*, **36**, 259–267.

McManus, M.E., Burgess, W.M., Veronese, M.E., Huggett, A., Quattrochi, L.C. and Tukey, R.H. (1990) Metabolism of 2-acetylaminofluorene and benzo(a)pyrene and activation of food-derived heterocyclic amine mutagens by human cytochromes P-450. *Cancer Res.*, **50**, 3367–3376.

Meyer, U.A., Skoda, R.C. and Zanger, U.M. (1990) The genetic polymorphism of debrisoquine/sparteine metabolism – molecular mechanisms. *Pharmacol. and Ther.*, **46**, 297–308.

Mori, H. and Nishikawa, A. (1998) Naturally occurring organosulphur compounds as potential anticarcinogens. In: *Nutrition and Chemical Toxicity*, C. Ioannides, ed. John Wiley, Chichester, pp. 285–299.

Mugford, C.A. and Kedderis, G.L. (1998) Sex-dependent metabolism of xenobiotics. *Drug Metab. Rev.*, **30**, 441–498.

Murray, M. and Reidy, G.L. (1990) Selectivity in the inhibition of mammalian cytochromes P-450 by chemical agents. *Pharm. Rev.*, **42**, 85–101.

Nebert, D.W. (1989) The Ah locus: genetic differences in toxicity, cancer, mutation and birth defects. *CRC Crit. Rev. in Toxicol.*, **20**, 153–174.

Nebert, D.W. (1991) Role of genetics and drug metabolism in human cancer risk. *Mutat. Res.*, **247**, 267–281.

Nims, R.W. and Lubet, R.A. (1996) The CYP2B family. In: *Cytochromes P450: Metabolic and Toxicological Aspects*, C. Ioannides, ed. CRC Press, Boca Raton, FL, pp. 135–160.

Oguri, K., Yamada, H. and Yoshimura, H. (1994) Regiochemistry of cytochrome P450 isozymes. *Ann. Rev. Pharmacol. and Toxicol.*, **34**, 251–279.

Okey, A.B. (1990) Enzyme induction in the P-450 system. *Pharmacol. and Ther.*, **45**, 241–298.

Parke, D.V., Ioannides, C. and Lewis, D.F.V. (1991) The role of cytochromes P450 in the detoxication and activation of drugs and other chemicals. *Canad. J. Physiol. and Pharmacol.*, **69**, 537–549.

Pasanen, M., Haaparanta, M., Sundin, M., Sivonen, P., Vahakangas, K., Raunio, H., Hines, R., Gustafsson, J-Å. and Pelkonen, O. (1990) Immuno-chemical and molecular biological studies on human placental cigarette smoke-inducible cytochrome P-450-dependent monooxygenase activities. *Toxicology*, **62**, 175–187.

Pasanen, M., Pellinen, P., Stenback, F., Juvonen, R.A., Raunio, H. and Pelkonen, O. (1995) The role of CYP enzymes in cocaine-induced liver damage. *Arch. Toxicol.*, **69**, 287–290.

Petruzzelli, S., Camus, A-M., Carozzi, L., Cherladucci, L., Rindi, M., Menconi, G., Angeletti, C.A., Ahotupa, M., Hietanen, E., Aitio, A., Saracci, R., Bartsch, H. and Giuntini, C. (1988) Long-lasting effects of tobacco smoking on pulmonary drug metabolizing enzymes: a case control study on lung cancer patients. *Cancer Res.*, **48**, 4695–4670.

Phillipson, C.E., Godden, P.M.M., Lum, P.Y., Ioannides, C. and Parke, D.V. (1984) Determination of cytochrome P-448 activity in biological tissues. *Biochem. J.*, **221**, 81–88.

Pirmohamed, M. and Park, B.K. (1996) Cytochromes P450 and immunotoxicity. In: *Cytochromes P450: Metabolic and Toxicological Aspects*, C. Ioannides, ed. CRC Press, Boca Raton, FL, pp. 329–354.

Poet, T.S., Brendel, K. and Halpert, J.R. (1994) Inactivation of cytochromes P450 2B protects against cocaine-mediated toxicity in rat liver slices. *Toxicol. and App. Pharmacol.*, **126**, 26–32.

Potter, T.D. and Coon, M.J. (1991) Cytochrome P450. Multiplicity of isoforms, substrates, and catalytic and regulatory mechanisms. *J. Biol. Chem.*, **266**, 13469–13472.

Puntarulo, S. and Cederbaum, A.I. (1998) Production of reactive oxygen species by microsomes enriched in specific human cytochrome P450 enzymes. *Free Radical Biol. and Med.*, **24**, 1324–1330.

Pykko, K., Tuimala, R., Aalto, L. and Perkio, T. (1991) Is aryl hydrocarbon hydroxylase activity a new prognostic indicator for breast cancer? *Br. J. of Cancer*, **63**, 596–600.

Rekka, E., Ayalogu, E.O., Lewis, D.F.V., Gibson, G.G. and Ioannides, C. (1994) Induction of hepatic microsomal CYP4A activity and of peroxisomal β-oxidation by two non-steroidal antiinflammatory drugs. *Arch. Toxicol.*, **68**, 73–78.

Rendic, S. and Di Carlo, F.J. (1997) Human cytochrome P450 enzymes: a status report summarizing their reactions, substrates, inducers, and inhibitors. *Drug Metab. Rev.*, **29**, 413–580.

Richardson, T.H. and Johnson, E.F. (1996) The CYP2C subfamily. In *Cytochromes P450: Metabolic and Toxicological Aspects*, C. Ioannides, ed. CRC Press, Boca Raton, FL, pp. 161–181.

Robertson, I.G.C., Serabjit-Singh, C., Croft, J.E. and Philpot, R.M. (1983) The relationship between increases in the hepatic content of hepatic cytochrome P-450 form 5, and in the metabolism of aromatic amines to mutagenic products following treatment of rabbits with phenobarbital. *Molec. Pharmacol.*, **24**, 156–162.

Ronis, M.J.J., Lindros, K.O. and Ingelman-Sundberg, M. (1996) The CYP2E subfamily. In: *Cytochromes P450: Metabolic and Toxicological Aspects*, C. Ioannides, ed. CRC Press, Boca Raton, FL, pp. 211–239.

Roy, D., Bernhardt, A., Strobel, H.W. and Liehr, J.G. (1992) Catalysis of the oxidation of steroid and stilbene estrogens to estrogen quinone metabolites by the β-naphthoflavone-inducible cytochrome P450IA family. *Arch. Biochem. and Biophys.*, **296**, 450–456.

Ryan, D.E. and Levin, W. (1990) Purification and characterization of hepatic cytochrome P-450. *Pharmacol. and Ther.*, **45**, 153–239.

Schuetz, J.D., Beach, D.L. and Guzelian, P.S. (1994) Selective expression of cytochrome P450 CYP3A mRNAs in embryonic and adult human liver. *Pharmacogenetics*, **4**, 11–20.

Service, R.F. (1998) New role for estrogen in cancer? *Science*, **279**, 1631–1633.

Sesardic, D., Boobis, A.R. and Davis, D.S. (1988) A form of cytochrome P-450 in man, orthologous to form d in the rat, catalyses the *O*-deethylation of phenacetin and is inducible by cigarette smoking. *Br. J. of Clin. Pharmacol.*, **26**, 363–372.

Shah, R.S., Oates, N.S., Idle, J.R., Smith, R.L. and Lockhart, J.D.F. (1982) Impaired oxidation of debrisoquine in patients with perhexiline neuropathy. *Br. Med. J.*, **284**, 295–299.

Shapiro, B.H., Agrawal, A.K. and Pampori, N.A. (1995) Gender differences in drug metabolism regulated by growth hormone. *Int. J. of Biochem. and Cell Biol.*, **27**, 9–20.

Shimada, T., Hayes, C.I., Yamazaki, H., Amin, S., Hecht, S.S., Guengerich, F.P. and Sutter, T.R. (1996) Activation of chemically diverse procarcinogens by human cytochrome P-450. *Cancer Res.*, **56**, 2979–2984.

Shimada, T., Iwasaki, M., Martin, M.V. and Guengerich, F.P. (1989) Human liver microsomal cytochrome P-450 enzymes involved in the bioactivation of procarcinogens detected by *umu* gene response in *Salmonella typhimurium* TA1535/pSK1002. *Cancer Res.*, **49**, 3218–3228.

Shimada, T., Nakamura, S-J., Imaoka, S. and Funae, Y. (1987) Genotoxic and mutagenic activation of aflatoxin B_1 by constitutive forms of cytochrome P-450 in rat liver microsomes. *Toxicol. and Appl. Pharmacol.*, **91**, 13–21.

Shimada, T., Yamazaki, H., Mimura, M., Inui, Y. and Guengerich, F.P. (1994) Inter-individual variations in human liver cytochrome P-450 enzymes involved in the oxidation of drugs: Studies with liver microsomes of 30 Japanese and 30 Caucasians. *J. Pharmacol. and Exp. Ther.*, **270**, 414–423.

Shou, M., Korzekwa, K.R., Brooks, E.N., Krauss, K.W., Gonzalez, F.J. and Gelboin, H.V. (1997) Role of human cytochrome P450 1A2 and 3A4 in the metabolic activation of estrone. *Carcinogenesis*, **18**, 207–214.

Shu, L. and Hollenberg, P.F. (1997) Alkylation of cellular macromolecules and target specificity of carcinogenic nitrosodialkylamines: metabolic activation by cytochromes P450 2B1 and 2E1. *Carcinogenesis*, **18**, 801–810.

Smith, T.J., Guo, Z., Gonzalez, F.J., Guengerich, F.P., Stoner, G.D. and Yang, C.S. (1992) Metabolism of 4-(methylnitrosamine)-*l*-(3-pyridyl)-*l*-butanone (NNK) in human lung and liver microsomes and in human cells containing cDNA-expressed cytochromes P-450. *Cancer Res.*, **52**, 1757–1763.

Sogawa, K. and Fujii-Kuriyama, Y. (1997) Ah receptor, a novel ligand-activated transcription factor. *J. of Biochem.*, **122**, 1075–1079.

Sonderfan, A.J., Arlotto, M.P. and Parkinson, A. (1989) Identification of the cytochrome P-450 isozymes responsible for testosterone oxidation in rat lung, kidney, and testis: evidence that cytochrome P-450a (P450IIA1) is the physiologically important testosterone 7 alpha-hydroxylase in rat testis. *Endocrinology*, **125**, 857–866.

Spence, J.D. (1997) Drug interactions with grapefruit: Whose responsibility is to warn the public? *Clin. Pharmacol. and Ther.*, **61**, 395–400.

Steele, C.M. and Ioannides, C. (1986) Induction of rat hepatic mixed function oxidases by aromatic amines and its relationship to their bioactivation to mutagens. *Mutat. Res.*, **162**, 41–46.

Steele, C.M., Masson, H.A., Battershill, J.M., Gibson, G.G. and Ioannides, C. (1983) Metabolic activation of paracetamol by highly purified forms of cytochrome P-450. *Res. Commun. in Chem. Path. and Pharmacol.*, **40**, 109–119.

Thornton-Manning, J.R., Nikula, K.J., Hotchkiss, J.A., Avila, K.J., Rohrbacher, K.D., Ding, X. and Dahl, A.R. (1997) Nasal cytochrome P450 2A: identification, regional localization, and metabolic activity toward hexamethylphosphoramide, a known nasal carcinogen. *Toxicol. and Appl. Pharmacol.*, **142**, 22–30.

Thornton-Manning, J.R., Ruangyuttikarn, W., Gonzalez, F.J. and Yost, G.R. (1991) Metabolic activation of the pneumotoxin, 3-methylindole by vaccinia-expressed cytochrome P450s. *Biochem. and Biophys. Res. Commun.*, **181**, 100–107.

Thummel, K.E. and Wilkinson, G.R. (1998) In vitro and in vivo drug interactions involving human CYP3A. *Ann. Rev. Pharmacol. and Toxicol.*, **38**, 389–430.

Verna, L., Whysner, J. and Williams, G.M. (1996) 2-Acetylaminfluorene mechanistic data and risk assessment: DNA reactivity, enhanced cell proliferation and tumor initiation. *Pharmacol. and Ther.*, **71**, 83–105.

Verschoyle, R.D., Philpot, R.M., Wolf, C.R. and Dinsdale, D. (1993) CYP4B1 activates 4-ipomeanol in the rat lung. *Toxicol. and Appl. Pharmacol.*, **123**, 193–198.

Walker, N.J., Gastel, J.A., Costa, L.T., Clark, G.C., Lucier, G.W. and Sutter, T.R. (1995) Rat CYP1B1: an adrenal cytochrome P450 that exhibits sex-dependent expression in livers and kidneys of TCDD-treated animals. *Carcinogenesis*, **16**, 1319–1327.

Wall, K.I., Gao, W., TeKoppele, J.M., Kwei, G.Y., Kauffman, F.C. and Thurman, R.G. (1991) The liver plays a central role in the mechanism of chemical carcinogenesis due to polycyclic aromatic hydrocarbons. *Carcinogenesis*, **12**, 783–786.

Wandel, C., Bocker, R., Bohrer, H., Browne, A., Rugheimer, E. and Martin, E. (1994) Midazolam is metabolized by at least three different cytochrome P450 enzymes. *Br. J. of Anaesth.*, **73**, 658–661.

Waxman, D.J., Lapenson, D.P., Aoyama, T., Gelboin, H.V., Gonzalez, F.J. and Korzekwa, K. (1991) Steroid hormone hydroxylase specificities of eleven cDNA-expressed human cytochrome P450s. *Arch. of Biochem. and Biophys.*, **290**, 160–166.

Westin, S., Tollet, P., Ström, A., Mode, A. and Gustafsson, J-Å. (1992) The role and mechanism of growth hormone in the regulation of sexually dimorphic cytochrome P450 enzymes in rat liver. *J. of Steroid Biochem. and Molecul. Biol.*, **43**, 1045–1053.

Wrighton, S.A. and Stevens, J.C. (1992) The human hepatic cytochromes P450 involved in drug metabolism. *Crit. Rev. in Toxicol.*, **22**, 1–21.

Yamazaki, H., Inui, H., Yun, C.H., Mimura, M., Guergerich, F.P. and Shimada, T. (1992) Cytochrome P4502E1 and 2A6 enzymes as major catalysts for metabolic activation of *N*-nitrosodialkylamines and tobacco-related nitrosamines in human liver microsomes. *Carcinogenesis*, **13**, 1789–1794.

Yamazaki, H., Shaw, P.M., Guengerich, F.P. and Shimada, T. (1998) Roles of cytochromes P450 1A2 and 3A4 in the oxidation of estradiol and estrone in human liver microsomes. *Chem. Res. in Toxicol.*, **11**, 659–665.

Yoshimoto, F., Masuda, S., Higashi, T., Nishii, T., Takamisawa, A., Tateishi, N. and Sakamoto, Y. (1985) Comparison of drug metabolizing activities in the livers of carcinogen-sensitive parent drugs and carcinogen-resistant descendants. *Cancer Res.*, **45**, 6155–6159.

Zhou, H.L., Anthony, L.B., Wood, A.J. and Wilkinson, G.R. (1990) Induction of polymorphic 4′-hydroxylation of *S*-mephenytoin by rifampicin. *Br. J. of Clin. Pharmacol.*, **30**, 471–475.

Zimmerman, H.J. and Maddrey, W.C. (1995) Acetaminophen (paracetamol) hepatotoxicity with regular intake of alcohol. Analysis of instances of therapeutic misadventure. *Hepatology*, **22**, 767–773.

5 Apoptosis Triggered by Free Radicals: Role in Human Diseases

GEORGE E. N. KASS,[1,4] **SEK C. CHOW**[2]
AND STEN ORRENIUS[3]

[1]*School of Biological Sciences, University of Surrey, Guildford, Surrey,* [2]*Centre for Mechanisms of Human Toxicity, University of Leicester, Leicester, UK,* [3]*Institute of Environmental Medicine, Karolinska Institutet, Stockholm, Sweden,* [4]*To whom correspondence should be addressed at the School of Biological Sciences, University of Surrey, Guildford, Surrey GU2 7XH, UK*

Introduction

Living organisms are unavoidably exposed to free radicals of various chemical and biological origins (Halliwell and Gutteridge 1999). The sources of these chemically highly reactive species range from physiological production, as a consequence of 'leaky' metabolic pathways such as the mitochondrial electron transport chain or from formation by specialised enzymes including nitric oxide synthase (NOS) and NADPH oxidase. Alternatively, they are produced accidentally (or intentionally) from exposure to ionising radiation, transition metals or drugs that are metabolised to free radical species or generate free radicals as by-products of their biotransformation.

The scientific and medical communities have witnessed over the past three decades a tremendous interest in the biological effects of free radicals, particularly those generated during conditions of oxidative stress. The interaction of free radicals with living organisms has traditionally focused on their cytotoxicity that results from their high chemical reactivity towards many cellular components (Halliwell and Gutteridge 1990; Kass *et al.* 1992b; Sies 1993; Halliwell 1994; Halliwell and Gutteridge 1999). The targets for free radicals include the cell's antioxidant and free-radical scavenging group of molecules (e.g. glutathione, ascorbate, vitamin E) as well as unsaturated fatty acids of cellular membrane lipids and phospholipids, nucleic acids and thiol and tyrosine moieties on proteins.

The mechanisms by which free radicals kill cells have been extensively studied over the past decades. The current consensus is that free radicals, by virtue of their direct or indirect interaction with oxygen to form reactive oxygen species and oxygen-based radicals such as superoxide anion (O_2^-), hydroxyl radical ($OH^\bullet$) and nitric oxide (NO) and peroxynitrite ($ONOO^-$), damage the target cell by inducing a condition of oxidative stress. Injury from free radicals can be either acute or chronic and may lead to cell death. However, recent evidence has also shown that levels of free radicals that are below the threshold for cytotoxicity can have very opposite effects such as inducing cell proliferation or differentiation, and this may result in diseases such as cancer (Kass 1997).

Cell death in tissues: apoptosis and programmed cell death

There is an obvious role for cell death in the shaping of organs and structures of the bodies of higher and lower organisms during embryogenesis, tissue development and also metamorphosis. The term *programmed cell death* was introduced by developmental biologists to describe the clear spatial and temporal patterns that control cell death (Ernst 1926; Glücksmann 1951; Lockshin and Williams 1965; Saunders 1966; Vaux and Korsmeyer 1999). With a few isolated exceptions, the role of cell death as an integral part of tissue homeostatic processes in the adult organism essentially remained unsuspected until the 1970s when Kerr, Wyllie and Currie coined the term *apoptosis* to show that a physiological form of cell death also occurs in the adult animal (Kerr *et al.* 1972). Since these initial observations, apoptosis has become a major focus of research in the fields of biology and medicine. Apoptosis is a fundamental and ubiquitous process that exists in all cell types in vertebrates (with the exception of mammalian red blood cells) and that is even found in plants and some unicellular organisms (Kerr *et al.* 1987; Arends and Wyllie 1991; Ellis *et al.* 1991; Rudin and Thompson 1997; Jacobson *et al.* 1997; Granville *et al.* 1998; Saini and Walker 1998).

Apoptosis regulates the physiological removal of cells during the entire lifespan of lower and higher organisms (Kerr *et al.* 1987; Arends and Wyllie 1991; Ellis *et al.* 1991; Rudin and Thompson 1997; Jacobson *et al.* 1997; Granville *et al.* 1998; Saini and Walker 1998). In this function, it forms an integral part of normal tissue homeostatic processes that control the steady-state number of cells of an organ, and should be regarded as the natural counterpart to mitosis. Apoptosis is also responsible for the decrease in the size of tissues such as endometrium, breast, liver and kidney in response to alterations in hormonal status or workload (Table 5.1). Another role of apoptosis is the destruction of cells that are potentially harmful to the body. For example, the removal of cancerous cells, cells infected with viruses and autoreactive T lymphocytes occurs through apoptosis. Failure of apoptosis

Table 5.1. Occurrence and roles of apoptosis in mammalian tissues

Embryogenesis (programmed cell death)
Removal of excess neurones
Removal of interdigital webs during limb development
Formation of coelomic cavity

Tissue homeostasis
Counteract proliferation (intestine, liver, kidney)
Regression of sexual organs

Defence mechanism
Killing of mutated and cancerous cells
Killing of virally infected cells
Killing autoreactive T lymphocytes

Ageing
Neutrophils, neurones

Table 5.2. Diseases resulting from inappropriate regulation of apoptosis

Examples of diseases caused by insufficient apoptosis	Examples of diseases caused by excessive apoptosis
Cancers	*Infections by viruses*
Hormone-dependent cancers (breast cancer, prostate cancer, liver cancer), follicular lymphomas	AIDS, hepatitis, diabetes
	Neurodegenerative diseases
	Parkinson's disease, Alzheimer's disease,
Autoimmune diseases	transmissible spongiform
Systemic lupus erythematosus	encephalopathies
Infections by viruses	*Immune-mediated diseases*
Herpes simplex, infectious mononucleosis	Diabetes, primary biliary cirrhosis, thyroiditis

to eliminate these cells may lead to the development of pathological conditions such as cancer or autoimmune diseases (Table 5.2). In contrast, apoptosis may also directly be responsible for certain diseases when inappropriately activated. Hence, conditions like AIDS, diabetes and neurodegenerative diseases, such as Parkinson's disease, Alzheimer's disease and amyotrophic lateral sclerosis, have all been attributed to an over-stimulation as well as inappropriate stimulation of apoptosis (Table 5.2) (Thompson 1995).

Free radicals count among the many different stimuli that can trigger cells to undergo apoptosis. Consequently, it is not surprising to find that free radicals have been causally related to human diseases that are commonly thought to arise from a dysregulation of apoptosis. In fact, many diseases where an involvement of free radicals has been suggested (such as neurodegenerative diseases, cancer, ageing, atherosclerosis, autoimmune diseases, endotoxic damage and drug-induced damage) (Halliwell and Gutteridge 1999), are also listed among those attributed to improper apoptosis (Thompson 1995). The aim of this review is to attempt to summarise our current understanding of the mechanisms by which free radicals induce apoptosis, and consequently, how free radicals may potentially be responsible for human diseases.

Apoptosis: morphological and biochemical considerations

Apoptosis was initially described by its morphological characteristics, which are summarised in Table 5.3. It should be pointed out that apoptosis is a continuum of interlinked rather than isolated events. Among these, several typical morphological features warrant to be highlighted. One early change in apoptosis is cell shrinkage. Concomitantly, marked alterations in cell shape occur with the loss of microvilli and the appearance of small protrusions on the cell surface that are commonly referred to as blebs (that yield a budding appearance to the cell) (Kerr *et al.* 1987; Arends and Wyllie 1991). In addition, a striking change in nuclear morphology, characterised by shrinkage of the nucleus and condensation of the nuclear chromatin

Table 5.3.   Morphological features of apoptosis

Cell shrinkage and detachment from neighbouring cells
Loss of microvilli and appearance of plasma membrane blebs
Margination of nuclear chromatin to nuclear envelope followed by condensation
Fragmentation of nucleus and cell into discrete units (apoptotic bodies)
Organelles (other than nucleus) remain intact, except for swelling of endoplasmic reticulum and some vacuolation

Biochemical features of apoptosis

Loss of plasma membrane asymmetry and appearance of phosphatidylserine on extracellular side
Expression of novel surface markers
Proteolysis of cellular constituents
DNA degradation to high molecular weight and oligonucleosomal-length fragments
Activation of transglutaminase

occurs. The normal appearance of the chromatin rapidly disappears in apoptosis as a result of the early margination of the chromatin towards the nuclear envelope and subsequent condensation to produce a doughnut-like appearance. The chromatin further condenses to a concave shape often resembling a half-moon. At this stage the DNA stains strongly with binding dyes giving it hyperfluorescent properties. Subsequently, and as a typically late event in apoptosis, the nuclear envelope breaks up, and the nucleus fragments to form discrete chromatin containing bodies called apoptotic bodies. Electron microscopic analysis of apoptotic cells has shown that cellular organelles other than the nucleus initially remain morphologically intact (Kerr et al. 1987; Arends and Wyllie 1991).

Apoptotic cells are rapidly phagocytosed by macrophages as well as by neighbouring cells, hence the late morphological changes in apoptosis are not always easily identifiable in vivo. The consequence of this rapid uptake and elimination of apoptotic cells in vivo is that the true extent of apoptosis occurring in a tissue can be completely underestimated. This is generally regarded as being one of the major reasons why the true importance of apoptosis in adult tissue homeostasis has so long failed to be recognised.

Numerous biochemical and molecular alterations have been described to occur during apoptosis. Early in apoptosis, the nuclear DNA is cleaved into distinct high molecular weight (HMW) fragments corresponding to >700 kilobase pairs (kbp), followed by further degradation to 200–250 and 50 kbp fragments (Oberhammer et al. 1993; Weis et al. 1995). Eventually, these are further degraded to oligonucleosomal-sized DNA fragments that produce a typical ladder appearance when resolved by agarose gel electrophoresis (Wyllie 1980; Arends and Wyllie 1991; Weis et al. 1995). The identity of the endonuclease(s) that cleave(s) DNA is still not completely established; however, it appears that the formation of HMW and oligonucleosomal-sized fragments can be mediated by separate enzymes (Sun and Cohen 1994; Zhivotovsky et al. 1994; Liu et al. 1997; Enari et al. 1998).

An additional major characteristic feature of apoptosis is the activation of a novel family of cysteine proteases that specifically cleave their substrates at a residue that is located next to an Asp residue (at P_1 position). The members of this family of proteases are called caspases and are related to the prototype enzyme, caspase-1, also known as interleukin-1β converting enzyme (Cohen 1997; Nicholson and Thornberry 1997; Thornberry and Lazebnik 1998; Nunez et al. 1998). At least 15 different caspases have been identified. Based on their tetrapeptide substrate specificity, which has been identified through a combinatorial approach, it is possible to distinguish between three groups of caspases based on their P_4 preferences (Nicholson and Thornberry 1997). Group I caspases prefer bulky hydrophobic residues at P_4, whereas groups II and III enzymes require Asp and branched amino acids at P_4, respectively. There is considerable evidence that members of group II caspases (in particular caspase-3, caspase-7 and caspase-2) are responsible for most of the features of apoptosis (except for nuclear lamin cleavage), whereas group III caspases (e.g. caspase-8, caspase-9, caspase-10) act upstream of the apoptotic pathways by processing and activating group II caspases. Thus, the cleavage of cytoskeletal elements by caspases appears to contribute to the changes in cell morphology. Likewise, the proteolysis of nuclear lamins has been linked to chromatin condensation, and the endonuclease causing the degradation of the nuclear DNA to oligonucleosomal-length fragments requires for its activation the proteolytic removal of an inhibitory protein. The classification of the caspases involved in apoptosis into two groups also fits a model that is rapidly receiving experimental support where the caspases are organised in a self-amplifying cascade to allow the rapid and complete execution of apoptosis. The role of group I caspases appears restricted to the proteolytic activation of pro-inflammatory cytokines such as pro-interleukin 1β. The reader is referred to recent reviews (Cohen 1997; Nicholson and Thornberry 1997; Thornberry and Lazebnik 1998; Nunez et al. 1998) for a comprehensive discussion on the role of caspases in apoptosis.

Additional biochemical characteristics of apoptosis include the loss of plasma membrane asymmetry with the appearance of phosphatidylserine (PS) on the extracellular side of the plasma membrane (Fadok et al. 1992). The latter event is believed to play a pivotal role in the recognition of apoptotic cells for removal by phagocytosis. Another important event in apoptosis is the activation of the enzyme transglutaminase (Fesus et al. 1987) that cross-links cytoplasmic proteins. This provides a rigid frame structure within apoptotic bodies, and functions to maintain their integrity and prevent the leakage of cellular constituents into the extracellular environment. Finally, there is evidence for changes in sphingolipid metabolism (Hannun 1996; Kolesnick and Kronke 1998) and calcium homeostasis (Kass and Orrenius 1999) during apoptosis. How the exposure of PS, activation of transglutaminase and changes in sphingolipid and calcium metabolism relate to the activation of caspases remains presently unclear.

Free radicals and apoptosis

Direct evidence for the induction of apoptosis by free radicals has been obtained from several studies. For instance, the formation of free radicals by ionising radiation leads to apoptosis in many cell types (Montagna and Wilson 1955; Hugon and Borgers 1966; Servomaa and Rytömaa 1990; Lennon et al. 1991; Zhivotovsky et al. 1993). Oxygen free radicals may also be generated by redox-cycling quinones and these compounds have been shown to trigger apoptosis in RINm5F insulinoma cells (Dypbukt *et al.* 1994) and FL5. 12 B lymphocyte precursor cells (Hockenbery *et al.* 1993). Likewise, the direct addition of the oxidant H_2O_2 to a range of cells, including endothelial cells, renal epithelial cells, oligodendrocytes and cells of haematological origin, causes cell death by apoptosis (Gramzinski *et al.* 1990; Lennon *et al.* 1991; Ueda and Shah 1992; Hockenbery *et al.* 1993; de Bono and Yang 1995; Marini *et al.* 1996; Gardner *et al.* 1997; Hampton and Orrenius 1997; Bhat and Zhang 1999).

Recent work has shown that NO, either through direct exposure of cells to a NO-generating compound (such as 3-morpholinosydnomine or *S*-nitrocysteine) or following endogenous formation by nitric oxide synthase (such as after exposure of neurones to glutamate receptor agonists), can trigger cells to apoptose. For example, excess NO production has been shown to be involved in glutamate-induced cell death of neuronal cultures (Dawson *et al.* 1991). Indeed, the exposure of neurones in culture to low concentrations of NO-generating compounds leads to the development of apoptosis (Bonfoco *et al.* 1995), and cortical cultures from neuronal NOS-deficient mice are resistant to glutamate neurotoxicity (Dawson *et al.* 1996). Nitric oxide has also been reported to induce apoptosis in non-neuronal cells, including macrophages (Albina *et al.* 1993; Sarih *et al.* 1993; Messmer *et al.* 1995; Shimaoka *et al.* 1995), pancreatic islet cells and β cell lines (Sarih *et al.* 1993; Ankarcrona *et al.* 1994; Kaneto *et al.* 1995; Mabley *et al.* 1997), human chondrocytes (Blanco *et al.* 1995) and mouse thymocytes (Fehsel *et al.* 1995). NO is a very reactive radical and under appropriate conditions can react with superoxide radical to form peroxynitrite ($ONOO^-$) which in turn has been found to induce apoptosis in several cell systems (Lin *et al.* 1995; Estevez *et al.* 1995).

In addition to the above cases where the direct exposure of cells to free radicals or free radical-generating systems leads to apoptosis, there is an even larger body of indirect evidence supporting the involvement of free radicals in apoptosis. A number of apoptotic signals such as tumour necrosis factor, although not a free radical itself, has been found to stimulate cellular free radical formation (Schulze-Osthoff *et al.* 1992; Shoji *et al.* 1995). The protective effect against most apoptotic stimuli afforded by a wide range of natural and synthetic antioxidants, transition metal ion chelators and spin traps provides further support for the critical role of free radicals in apoptosis (Wong *et al.* 1989; Matsuda *et al.* 1991; Chang *et al.* 1992; Wolfe *et al.* 1994; Buttke and Sandstrom 1994; Mayer and Noble 1994; Slater *et al.* 1995; Bustamante *et al.* 1995). Interestingly, we and others have found that the process of apoptosis itself is coupled to the loss of endogenous glutathione through activation of a GSH efflux mechanism (Ghibelli *et al.* 1995; Van den Dobbelsteen *et al.* 1996); this in turn leads to enhanced cellular free radical formation even though the primary apoptotic trigger is devoid of any redox or radical forming activity. Hence free radicals may be linked to apoptosis in a more universal way than previously thought.

Free radicals and cell death: apoptosis versus necrosis

The ability of free radicals to induce lethal cell injury is undisputed. However, there has been considerable confusion in the literature about whether free radicals kill cells by apoptosis or by necrosis. Necrosis, as opposed to apoptosis, results from severe damage to cells leading to rapid energy loss, failure of ionic pumps and a generalised collapse of cellular homeostatic mechanisms (Arends and Wyllie 1991; Trump *et al*. 1997). The plasma membrane loses its barrier function and as a consequence there is a release of intracellular contents, which in turn provokes an inflammatory reaction and further injury to the surrounding tissue.

Necrosis and apoptosis are often considered as two fundamentally opposite forms by which a cell can die. However, it now appears that we are dealing with a continuum of overlapping mechanisms of cell death (Raffray and Cohen 1997; Lipton and Nicotera 1998; Kass and Orrenius 1999). For instance, several important features such as plasma membrane blebbing and mitochondrial damage are common to both apoptosis and necrosis. Also, the anti-apoptotic gene product Bcl-2 has been found to protect cells from both apoptosis and necrosis (Strasser *et al*. 1991; Kane *et al*. 1993; Kane *et al*. 1995; Shimizu *et al*. 1996). Moreover, when apoptosis occurs at an excessive level, the dying cells can no longer be removed in an orderly fashion by phagocytosis, leaving the cells to degenerate into 'secondary' necrosis and inflammation to occur (Ogasawara *et al*. 1993; Majno and Joris 1995; Kass and Jones 1999).

It is evident that these confounding factors need to be taken into account when examining the literature on free radical-induced cytotoxicity. Several recent studies have systematically examined the conditions that lead to apoptosis as compared to those that induce necrosis; a pattern has emerged whereby low levels of free radical generation or a relatively brief exposure favours apoptosis over necrosis, whereas high levels or prolonged exposure times predominantly leads to necrosis. In other words, the severity of the insult determines the fate of the cell. This phenomenon has been observed in RINmF5 insulinoma cells exposed to oxygen free radical-generating quinones (Dypbukt *et al*. 1994), L929 fibroblasts and Jurkat T lymphocytes treated with H_2O_2 (Gardner *et al*. 1997; Hampton and Orrenius 1997), lymphocytes exposed to γ-irradiation (Sellins and Cohen 1987), as well as Jurkat T lymphocytes, RAW 264.7 macrophages and cortical neuronal cultures exposed to NO-generating compounds (Bonfoco *et al*. 1995;

Ankarcrona *et al*. 1995; Messmer and Brüne 1996; Leist *et al*. 1999) and PC12 cells exposed to peroxynitrite (Estevez *et al*. 1995).

If free radicals can induce either apoptosis or necrosis depending on the severity of the insult, what are then the factors that determine the type of death of a cell? As described in the next sections, the identity of some of these determining factors has recently emerged alongside with a better understanding of the molecular mechanisms involved in apoptosis.

Mechanisms of free radical-induced apoptosis

Mammalian cells can undergo apoptosis through two primary pathways (Figure 5.1). One is through the activation of so-called death receptors (CD95, TNFR1, DR-3, DR-4, DR-5, etc.) that are functionally coupled to the execution caspase-3 and possibly caspase-7 through caspase-8 (Slee *et al*. 1996; Jones *et al*. 1998; Peter and Krammer 1998; Wallach *et al*. 1998; Ashkenazi and Dixit 1999). More recently, it has become apparent that mitochondria are responsible for a second pathway of apoptosis. Although this pathway similarly activates the execution caspase-3, the underlying mechanism does not involve a cell surface receptor or caspase-8. Rather, the apoptotic signal is relayed by the mitochondria through the release of cytochrome *c*, a small protein that is located in the intermembrane space of mitochondria and that is well known as an integral component of the mitochondrial electron transport chain (Krippner *et al*. 1996; Liu *et al*. 1996; Kluck *et al*. 1997; Yang *et al*. 1997; Zhivotovsky *et al*. 1998). In the cytosol, cytochrome *c* induces the formation of a caspase-activating complex in the presence of dATP (Li *et al*. 1997; Hampton *et al*. 1998) that comprises pro-caspase-9 bound to Apaf-1 (one of the mammalian homologues of the *C. elegans* CED-4 protein) through a homophilic interaction involving caspase recruitment domain (CARD) motifs. The proteolytic activation of the proform of caspase-9 to its active form occurs through an intrinsic mechanism and is followed by the downstream proteolytic activation of pro-caspase-3 and pro-caspase-7 (Pan *et al*. 1998). Recently, we have found that the microinjection of cytochrome *c* into cells is sufficient to induce apoptosis (Zhivotovsky *et al*. 1998).

Not unexpectedly, the induction of apoptosis by free radicals has been found to require the processing and activation of caspases. For instance, the generation of free radicals as a result of exposure of Jurkat T lymphocytes to conditions of oxidative stress was found to lead to the activation of the execution caspase-3 (Hampton and Orrenius 1997; Ueda *et al* 1998; Stridh *et al*. 1998). The subsequent induction of apoptosis was dependent on this activation as assessed by the ability of caspase inhibitory peptides to antagonise apoptosis (Stridh *et al*. 1998). Likewise, H9c2 cardiac muscle cells

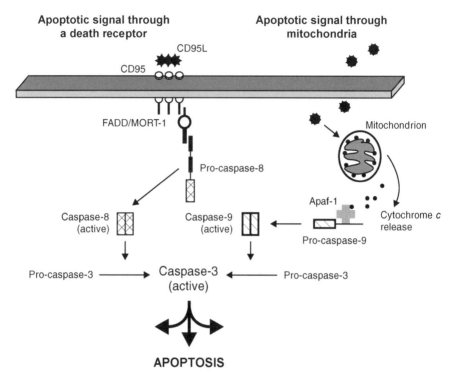

Figure 5.1. Two main pathways leading to apoptosis have been identified, one resulting from the activation of so-called death receptors, the other involving mitochondria. Death receptors such as CD95 (Fas, APO-1) or tumour necrosis factor receptor 1, trigger a cascade of killer proteases known as caspases that become activated with the help of relay proteins such as FADD/MORT-1. Alternatively, the apoptotic signal may target mitochondria resulting in the release of the electron carrier cytochrome c from the intermembrane space. Recent studies have shown that once in the cytosol cytochrome c binds a number of docking proteins such as Apaf-1 as a result of which pro-caspase-9 is cleaved to its active form, caspase-9. The convergence point of both apoptotic pathways is caspase-3 (and possibly caspase-7) whose activation appears to be essential for the demise of the cell committed to apoptosis. The mitochondrial may provide an amplification step for the efficient activation of caspases through the death receptor pathway

could be triggered to apoptose upon exposure to the oxygen free radical generator menadione as a result of the activation of caspases (Turner *et al.* 1998). In another report, the formation of radicals following UVA irradiation and apoptosis of HL60 cells also involved caspases (Tada-Oikawa *et al.* 1998).

Recent studies have shown that the activation of caspases during the apoptotic response to free radicals is preceded by a translocation of

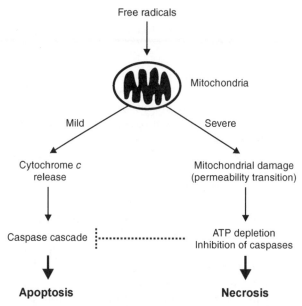

Free radicals

Mitochondria

Mild   Severe

Cytochrome *c*
release

Mitochondrial damage
(permeability transition)

Caspase cascade ⋮·····················

ATP depletion
Inhibition of caspases

Apoptosis   **Necrosis**

Figure 5.2. Interaction of free radicals with mitochondria. Current evidence suggests that mitochondria not only represent a convergence point for the action of free radicals but also the nature and extent of damage inflicted on mitochondria will decide whether cell death occurs by necrosis or apoptosis. Mild insult will lead to apoptosis through cytochrome *c* release, whereas severe injury will induce complete permeability transition that will compromise the mitochondrial functions and energy production. The ensuing loss of ATP will contribute to the death of the cell by necrosis through the inhibition of the apoptotic pathway

cytochrome *c* from mitochondria to the cytosol. This phenomenon has been observed as a result of the direct exposure of cells to reactive oxygen species or to free radical generating systems. For example, Jurkat cells treated with H_2O_2 (Stridh *et al.* 1998) or diamide (Ueda *et al.* 1998) respond by releasing cytochrome *c* from mitochondria prior to caspase-3 activation. The induction of apoptosis in a range of cell lines by 1,10-phenanthroline (Duan *et al.* 1999), 7-ketocholesterol (Lizard *et al.* 1998) or arsenite (Chen *et al.* 1998) involves the translocation of cytochrome *c* to the cytosol. NO-induced apoptosis is also characterised by cytochrome *c* release and caspase activation (Ushmorov *et al.* 1999; Uehara *et al.* 1999).

Taken together, the currently available evidence clearly suggests that the ability of free radicals to induce apoptosis in their target cells is dependent on the mitochondrial pathway of apoptosis, which involves the release of cytochrome *c* from mitochondria followed by activation of the caspase cascade. What is less clear is how the mitochondrial apoptotic pathway becomes activated by free radicals.

Mitochondria as molecular convergence point for free radical-induced cell death: apoptosis versus necrosis

Current evidence clearly points towards mitochondria as critical targets required for the induction of cell death by free radicals (Figure 5.2). Free radicals have been known for many years for their ability to cause mitochondrial damage, particularly through activation of a pathway known as permeability transition (Kass *et al.* 1992a; Lemasters *et al.* 1997). Mitochondrial permeability transition involves the opening of a pore (the permeability transition pore, PTP) that is made of a large multiprotein complex comprising among others the voltage-dependent anion channel (VDAC, porin), adenine nucleotide translocator (ANT), mitochondrial cyclophilin D, the peripheral benzodiazepine receptor, hexokinase, creatine kinase and also the proapoptotic gene product Bax (Zoratti and Szabò 1995; Bernardi and Petronilli 1996; Beutner *et al.* 1997). The pore complex has been localised to the contact sites between the inner and outer mitochondrial membranes. However, the molecular details of the association of the different components of the pore are still unclear although the conditions that trigger pore opening have been well worked out. The pore behaves as a voltage-operated channel that becomes activated by high matrix Ca^{2+}, oxidative stress, pyridine nucleotide oxidation, thiol oxidation and low transmembrane potential. Initially, rapid and stochastic opening and closing of the pore is observed (Huser *et al.* 1998; Bathori *et al.* 1998). This, however, soon develops into persistent pore opening, allowing low molecular weight molecules ($M_r < 1500$) to escape from the mitochondrial matrix (Weis *et al.* 1994). At this stage the opening of the pore is still reversible by agents such as the cyclophilin D ligand cyclosporin A (Weis *et al.* 1994). However, the molecular events that regulate the transition from an initially voltage-dependent anion channel to a megachannel remains obscure.

Based on knowledge gained from our studies performed on isolated mitochondria, free radicals can induce pore opening in one or more of several ways. It could occur through direct interaction of Ca^{2+} with the metal binding site of the pore (Weis *et al.* 1994). Indeed, free radicals are well known to cause severe perturbation of cellular Ca^{2+} homeostasis leading to a non-physiological increase in the concentration of Ca^{2+} in the cytosol Ca^{2+} (Jewell *et al.* 1982; Nicotera *et al.* 1986; Orrenius *et al.* 1992; Kass *et al.* 1992b; Trump and Berezesky 1996). Increases in cytosolic Ca^{2+} concentration are rapidly reflected by a similar increase in mitochondrial Ca^{2+} content (Juedes *et al.* 1992; Kass *et al.* 1992a). In addition, direct oxidation of critical thiol groups (Bernardi and Petronilli 1996) and pyridine nucleotides (Richter and Kass 1991; Weis *et al.* 1994), inhibition of respiration by NO through Ca^{2+} activation of the mitochondrial NO synthase (Ghafourifar and Richter 1997), oxidative stress imposed by the loss of glutathione during apoptosis (Ghibelli *et al.* 1995; Van den Dobbelsteen *et al.* 1996) as well as superoxide radical production triggered as a result of cytochrome *c* loss (Cai and Jones 1998) may also contribute to permeability transition.

In vitro studies have convincingly demonstrated that mitochondrial pore opening and permeability transition leads to cytochrome *c* release, probably as a result of the disruption of the outer membrane by the swelling associated with permeability transition (Kantrow and Piantadosi 1997; Petit *et al.* 1998). Yet, other reports have shown that cytochrome *c* is mobilised from mitochondria prior to, or in the complete absence of, permeability transition (Yang *et al.* 1997; Bossy-Wetzel *et al.* 1998).

One of the effectors of cytochrome *c* release in apoptosis may be Bax, a member of the pro-apoptotic subgroup of the Bcl-2 superfamily (Jürgensmeier *et al.* 1998; Pastorino *et al.* 1998). Bax has been proposed to directly cooperate with ANT to induce PTP opening (Marzo *et al.* 1998). Cytochrome *c* release would occur as a result of the disruption of the mitochondrial outer membrane caused by the swelling of the matrix from permeability transition. More recently, Tsujimoto and co-workers (Shimizu *et al.* 1999) have presented evidence for a radically different model for cytochrome *c* release where Bax or its homologue, Bak, interact with the mitochondrial outer membrane channel VDAC, altering its conductivity properties to form a megachannel. This allows cytochrome *c* to leave the intermembrane space and to interact with procaspase-9 through Apaf-1 and dATP. In their model, cytochrome *c* release (and hence activation of the execution caspases and apoptosis) may occur in the absence of mitochondrial permeability transition. Bcl-2 and Bcl-x_L prevent cytochrome *c* release, possibly through its direct interaction with Bax (Yang *et al.* 1997) or through closing VDAC by binding to it directly (Shimizu *et al.* 1999). In addition, a caspase-8-mediated cleavage product of the BH3-domain-containing Bid has been suggested to mediate the release of cytochrome *c* from mitochondria, also in the absence of permeability transition (Li *et al.* 1998; Luo *et al.* 1998).

The issue of mitochondrial permeability transition and hence cytochrome *c* release is critical to our understanding of how free radicals induce cell death. Indeed, we and others have previously reported that mitochondrial permeability transition induces rapid and severe swelling in cells and a complete loss of cellular energy (Kass *et al.* 1992a; Trump *et al.* 1997; Lemasters *et al.* 1997), all of these changes being more consistent with necrosis rather than apoptosis. Indeed, recent work has demonstrated that the apoptotic programme necessitates cellular ATP (Leist *et al.* 1997; Tsujimoto 1997), and it has been suggested that ATP acts as a major decision switch between apoptosis and necrosis (Figure 5.2). Using the CD95 (also known as Fas or APO-1) model for apoptosis these authors have demonstrated that a decrease in cellular ATP content induced through inhibition of the mitochondrial ATP synthase or through exposure to NO-generating compounds converts an apoptotic signal into a necrotic one. Likewise, Jurkat T lymphocytes exposed to the NO donors S-nitrosoglutathione, S-nitroso-N-acetylpenicillamine or spermine-NO were killed by apoptosis only in glucose-containing medium, whereas in the absence of this source of energy (and hence low cellular ATP levels) the same treatment induced necrosis (Leist *et al.* 1999). Two ATP requiring targets have been identified for their involvement in apoptosis, the mitochondrial ATP synthase and ANT (Marzo *et al.* 1998; Matsuyama *et al.* 1998). Furthermore, our own results have shown that ATP is necessary for nuclear chromatin condensation to occur during apoptosis (Kass *et al.* 1996).

Caspases as targets for free radicals in necrosis

In addition to cellular ATP levels as a regulator of apoptotic cell death, we and others have recently shown that a direct inactivation of caspases by free radicals can prevent apoptosis and favour cell death by necrosis. Caspases fall in the category of cysteine proteases, and consequently are very sensitive to inhibition through covalent and oxidative modification of one or more critical cysteine residue(s). Disulphides such as disulphiram and GSSG (in the form of a GSH:GSSG redox buffer) can readily inactivate caspases *in vitro* (Nobel *et al.* 1997b). This redox-based inhibition of caspases is responsible for the anti-apoptotic properties of dithiocarbamates (Nobel *et al.* 1997a). In cells exposed to levels of free radicals or oxidative stress that induce necrosis instead of apoptosis, no activation of caspase-3 occurred (Hampton and Orrenius 1997; Ueda *et al.* 1998; Leist *et al.* 1999; Samali *et al.* 1999), even though the translocation of cytochrome *c* to the cytosol was observed (Ueda *et al.* 1998; Leist *et al.* 1999; Samali *et al.* 1999). Experiments were carried out to examine the effects of free radicals on apoptosis triggered by the death receptor CD95. In the presence of free radicals and conditions of oxidative stress, apoptosis was severely impaired and cell death switched to necrosis (Hampton and Orrenius 1997; Leist *et al.* 1999; Samali *et al.* 1999). The transition to necrosis correlated with an inhibition of caspase activation by the apoptogenic signal (Hampton and Orrenius 1997; Leist *et al.* 1999; Samali *et al.* 1999). Closer examination of the mechanism of caspase inactivation by NO under conditions where necrosis is induced has suggested three sites of action. One is the S-nitrosylation of a critical cysteine residue in caspase-3 (and probably other caspases) by NO (Mohr *et al.* 1997; Tenneti *et al.* 1997; Li *et al.* 1997; Melino *et al.* 1997; Dimmeler *et al.* 1997). Another target is the release of cytochrome *c* that is delayed by NO (Leist *et al.* 1999). Finally, the processing of the pro-forms of caspase-3, caspase-7 and caspase-8 is impaired irrespective of cytochrome *c* release (Leist *et al.* 1999). Both the lack of caspase activation and cytochrome *c* release have been linked to a depletion of cellular ATP (Leist *et al.* 1999).

The inhibition of caspases by free radicals and pro-oxidants may be important in the modulation of apoptosis in various pathological and possibly physiological states. For instance, caspase activity is suppressed in activated neutrophils through endogenous pro-oxidant production (Fadeel *et al.* 1998), and this may have implications on how these cells are being cleared from circulation, as well as on the fate of bystander cells that are exposed to high levels of free radicals.

Conclusions and perspectives

Apoptosis plays an integral part of fundamental processes such as organogenesis, tissue homeostasis and cellular defence mechanisms. Although extremely tightly controlled, it is clear now that any interference with these control mechanisms, whether through genetic regulation or through the influence of environmental factors, including environmental pollutants, natural toxins or medicines, can have devastating effects on the functioning of the body. Among these agents, free radicals play a critical role in causing apoptosis and their role in apoptosis-related diseases is well accepted. If cells become overwhelmed by free radicals, as in some forms of acute injury, the response switches from apoptosis to necrosis, partly by overwhelming the cell's homeostatic processes but also partly by selectively downregulating the apoptotic pathway.

Currently, a great deal of effort is spent on selectively targeting apoptosis for therapeutic intervention and it is expected that the coming decade will provide the medical profession with a whole new range of tools to combat diseases. Hence, there may be a great deal of scope in targeting those diseases whose aetiology involves free radicals and apoptosis. Furthermore, as we begin to understand the events that alter an apoptotic signal into one that leads to necrosis, new targets enabling the switching between the two modes of cell death may emerge. Most of the efforts are geared towards controlling apoptosis; yet, we must not forget that diseases like cancer have evolved partly through a selective inactivation of the apoptotic machinery. In certain cases, a strategy aimed at killing cells by necrosis may be more effective, even though the trade-off would be inflammation and more widespread tissue destruction.

References

Albina, J.E., Cui, S., Mateo, R.B. and Reichner, J.S. (1993) Nitric oxide-mediated apoptosis in murine peritoneal macrophages. *J. Immunol.*, **150**, 5080–5085.

Ankarcrona, M., Dypbukt, J.M., Bonfoco, E., Zhivotovsky, B., Orrenius, S., Lipton, S.A. and Nicotera, P. (1995) Glutamate-induced neuronal death: a succession of necrosis or apoptosis depending on mitochondrial function. *Neuron*, **15**, 961–973.

Ankarcrona, M., Dypbukt, J.M., Brüne, B. and Nicotera, P. (1994) Interleukin 1β-induced nitric oxide production activates apoptosis in pancreatic RINm5F cells. *Exp. Cell Res.*, **213**, 172–177.

Arends, M.J. and Wyllie, A.H. (1991) Apoptosis: Mechanisms and roles in pathology. *Int. Rev. Exp. Pathol.*, **32**, 223–254.

Ashkenazi, A. and Dixit, V.M. (1999) Apoptosis control by death and decoy receptors. *Current Opinion Cell Biol.*, **11**, 255–260.

Bathori, G., Szabo, I., Schmehl, I., Tombola, F., DePinto, V. and Zoratti, M. (1998) Novel aspects of the electrophysiology of mitochondrial porin. *Biochem. Biophys. Res. Commun.*, **243**, 258–263.

Bernardi, P. and Petronilli, V. (1996) The permeability transition pore as a mitochondrial calcium-release channel – a critical appraisal. *J. Bioenerg. Biomembr.*, **28**, 131–138.

Beutner, G., Ruck, A., Riede, B. and Brdiczka, D. (1997) Complexes between hexokinase, mitochondrial porin and adenylate translocator in brain: regulation of hexokinase, oxidative phosphorylation and permeability transition pore. *Biochem. Soc. Trans.*, **25**, 151–157.

Bhat, N.R. and Zhang, P. (1999) Hydrogen peroxide activation of multiple mitogen-activated protein kinases in an oligodendrocyte cell line: Role of extracellular signal-regulated kinase in hydrogen peroxide-induced cell death. *J. Neurochem.*, **72**, 112–119.

Blanco, F.J., Ochs, R.L., Schwarz, H. and Lotz, M. (1995) Chondrocyte apoptosis induced by nitric oxide. *Am. J. Pathol.*, **146**, 75–85.

Bonfoco, E., Krainc, D., Ankarcrona, M., Nicotera, P. and Lipton, S.A. (1995) Apoptosis and necrosis: Two distinct events induced, respectively, by mild and intense insults with N-methyl-D-aspartate or nitric oxide/superoxide in cortical cell cultures. *Proc. Natl. Acad. Sci. US*, **92**, 7162–7166.

Bossy-Wetzel, E., Newmeyer, D.D. and Green, D.R. (1998) Mitochondrial cytochrome *c* release in apoptosis occurs upstream of DEVD-specific caspase activation and independently of mitochondrial transmembrane depolarization. *EMBO J.*, **17**, 37–49.

Bustamante, J., Slater, A.F.G. and Orrenius, S. (1995) Antioxidant inhibition of thymocyte apoptosis by dihydrolipoic acid. *Free Radical Biol. Med.*, **19**, 339–347.

Buttke, T.M. and Sandstrom, P.A. (1994) Oxidative stress as a mediator of apoptosis. *Immunol. Today*, **15**, 7–10.

Cai, J. and Jones, D.P. (1998) Superoxide in apoptosis: Mitochondrial generation triggered by cytochrome *c* loss. *J. Biol. Chem.*, **273**, 11401–11404.

Chang, D.J., Ringold, G.M. and Heller, R.A. (1992) Cell killing and induction of manganous superoxide dismutase by tumour necrosis factor α is mediated by lipoxygenase metabolites of arachidonic acid. *Biochem. Biophys. Res. Commun.*, **188**, 538–546.

Chen, Y.C., Lin-Shiau, S.Y. and Lin, J.K. (1998) Involvement of reactive oxygen species and caspase 3 activation in arsenite-induced apoptosis. *J. Cell. Physiol.*, **177**, 324–333.

Cohen, G.M. (1997) Caspases: the executioners of apoptosis. *Biochem. J.*, **326**, 1–16.

Dawson, V.L., Dawson, T.M., London, E.D., Bredt, D.S. and Snyder, S.H. (1991) Nitric oxide mediates glutamate neurotoxicity in primary cortical cultures. *Proc. Natl. Acad. Sci. US*, **88**, 6368–6371.

Dawson, V.L., Kizushi, V.M., Huang, P.L., Snyder, S.H. and Dawson, T.M. (1996) Resistance to neurotoxicity in cortical cultures from neuronal nitric oxide synthase-deficient mice. *J. Neurosci.*, **16**, 2479–2487.

De Bono, D.P. and Yang, W.D. (1995) Exposure to low concentrations of hydrogen peroxide causes delayed endothelial cell death and inhibits proliferation of surviving cells. *Atherosclerosis*, **114**, 235–245.

Dimmeler, S., Haendeler, J., Nehls, M. and Zeiher, A.M. (1997) Suppression of apoptosis by nitric oxide via inhibition of interleukin-1β-converting enzyme (ICE)-like and cysteine protease protein (CPP)-32-like proteases. *J. Exp. Med.*, **185**, 601–607.

Duan, H.J., Wang, Y.L., Aviram, M., Swaroop, M., Loo, J.A., Bian, J.H., Tian, Y., Mueller, T., Bisgaier, C.L. and Sun, Y. (1999) SAG, a novel zinc RING finger protein that protects cells from apoptosis induced by redox agents. *Mol. Cell Biol.*, **19**, 3145–3155.

Dypbukt, J.M., Ankarcrona, M., Burkitt, M., Sjöholm, Å., Ström, K., Orrenius, S. and Nicotera, P. (1994) Different prooxidant levels stimulate growth, trigger apoptosis, or produce necrosis of insulin-secreting RINm5F cells: The role of intracellular polyamines. *J. Biol. Chem.*, **269**, 30553–30560.

Ellis, R.E., Yuan, J. and Horvitz, H.R. (1991) Mechanisms and functions of cell death. *Annu. Rev. Cell Biol.*, **7**, 663–698.

Enari, M., Sakahira, H., Yokoyama, H., Okawa, K., Iwamatsu, A. and Nagata, S. (1998) A caspase-activated DNase that degrades DNA during apoptosis, and its inhibitor ICAD. *Nature*, **391**, 43–50.

Ernst, M. (1926) Über Untergang von Zellen während der normalen Entwicklung bei Wirbeltieren. *Z. Anat. Entwicklungsgesch.*, **79**, 228–262.

Estevez, A.G., Radi, R., Barbeito, L., Shin, J.T., Thompson, J.A. and Beckman, J.S. (1995) Peroxynitrite-induced cytotoxicity in PC12 cells: Evidence for an apoptotic mechanism differentially modulated by neurotrophic factors. *J. Neurochem.*, **65**, 1543–1550.

Fadeel, B., Åhlin, A., Henter, J.-I., Orrenius, S. and Hampton, M.B. (1998) Involvement of caspases in neutrophil apoptosis: Regulation by reactive oxygen species. *Blood*, **12**, 4808–4818.

Fadok, V.A., Voelker, D.R., Campbell, P.A., Cohen, J.J., Bratton, D.L. and Henson, P.M. (1992) Exposure of phosphatidylserine on the surface of apoptotic lymphocytes triggers specific recognition and removal by macrophages. *J. Immunol.*, **148**, 2207–2216.

Fehsel, K., Kröncke, K.D., Meyer, K.L., Huber, H., Wahn, V. and Kolb-Bachofen, V. (1995) Nitric-oxide induces apoptosis in mouse thymocytes. *J. Immunol.*, **155**, 2858–2865.

Fesus, L., Thomazy, V. and Falus, A. (1987) Induction and activation of tissue transglutaminase during programmed cell death. *FEBS Lett.*, **224**, 104–108.

Gardner, A.M., Xu, F.-H., Fady, C., Jacoby, F.J., Duffey, D.C., Tu, Y. and Lichtenstein, A. (1997) Apoptotic vs. nonapoptotic cytotoxicity induced by hydrogen peroxide. *Free Radical Biol. Med.*, **22**, 73–83.

Ghafourifar, P. and Richter, C. (1997) Nitric oxide synthase activity in mitochondria. *FEBS Lett.*, **418**, 291–296.

Ghibelli, L., Coppola, S., Rotilio, G., Lafavia, E., Maresca, V. and Ciriolo, M.R. (1995) Nonoxidative loss of glutathione in apoptosis via GSH extrusion. *Biochem. Biophys. Res. Commun.*, **216**, 313–320.

Glücksmann, A. (1951) Cell death in normal vertebrate ontogeny. *Biol. Rev.*, **26**, 59–86.

Gramzinski, R.A., Parchment, R.E. and Pierce, G.B. (1990) Evidence linking programmed cell death in the blastocyst to polyamine oxidation. *Differentiation*, **43**, 59–65.

Granville, D.J., Carthy, C.M., Hunt, D.C. and McManus, B.M. (1998) Apoptosis: Molecular aspects of cell death and disease. *Lab. Invest.*, **78**, 893–913.

Halliwell, B. (1994) Free radicals, antioxidants, and human disease: curiosity, cause, or consequence? *Lancet*, **344**, 721–724.

Halliwell, B. and Gutteridge, J.M.C. (1990) Role of free radicals and catalytic metal ions in human disease: An overview. *Methods Enzymol.*, **186**, 1–85.

Halliwell, B. and Gutteridge, J.M.C. (1999) *Free Radicals in Biology and Medicine*. Oxford University Press, Oxford.

Hampton, M.B. and Orrenius, S. (1997) Dual regulation of caspase activity by hydrogen peroxide: implications for apoptosis. *FEBS Lett.*, **414**, 552–556.

Hampton, M.B., Zhivotovsky, B., Slater, A.F.G., Burgess, D.H. and Orrenius, S. (1998) Importance of the redox state of cytochrome c during caspase activation in cytosolic extracts. *Biochem. J.*, **329**, 95–99.

Hannun, Y.A. (1996) Functions of ceramide in coordinating cellular responses to stress. *Science*, **274**, 1855–1859.

Hockenbery, D.M., Oltvai, Z.N., Yin, X.M., Milliman, C.L. and Korsmeyer, S.J. (1993) Bcl-2 functions in an antioxidant pathway to prevent apoptosis. *Cell*, **75**, 241–251.

Hugon, J. and Borgers, M. (1966) Ultrastructural and cytochemical studies on karyolytic bodies in the epithelium of the duodenal crypts of whole body X-irradiated mice. *Lab. Invest.*, **15**, 1528–1543.

Huser, J., Rechenmacher, C.E. and Blatter, L.A. (1998) Imaging the permeability pore transition in single mitochondria. *Biophys. J.*, **74**, 2129–2137.

Jacobson, M.D., Weil, M. and Raff, M.C. (1997) Programmed cell death in animal development. *Cell*, **88**, 347–354.

Jewell, S.A., Bellomo, G., Thorn, P., Orrenius, S. and Smith, M.T. (1982) Bleb formation in hepatocytes during drug metabolism is caused by disturbances in thiol and calcium ion homeostasis. *Science*, **217**, 1257–1259.

Jones, R.A., Johnson, V.L., Buck, N.R., Dobrota, M., Hinton, R.H., Chow, S.C. and Kass, G.E.N. (1998) Fas-mediated apoptosis in mouse hepatocytes involves the processing and activation of caspases. *Hepatology*, **27**, 1632–1642.

Juedes, M.J., Kass, G.E.N. and Orrenius, S. (1992) M-iodobenzylguanidine increases the mitochondrial Ca^{2+} pool in isolated hepatocytes. *FEBS Lett.*, **313**, 39–42.

Jürgensmeier, J.M., Xie, Z., Deveraux, Q., Ellerby, L., Bredesen, D.E. and Reed. J.C. (1998) Bax directly induces release of cytochrome c from isolated mitochondria. *Proc. Natl. Acad. Sci. US*, **95**, 4997–5002.

Kane, D.J., Ord, T., Anton, R. and Bredesen, D.E. (1995) Expression of bcl-2 inhibits necrotic neural cell-death. *J. Neurosci. Res.*, **40**, 269–275.

Kane, D.J., Sarafian, T.A., Anton, R., Hahn, H., Gralla, E.B., Valentine, J.S., Ord, T. and Bredesen, D.E. (1993) Bcl-2 inhibition of neural death: Decreased generation of reactive oxygen species. *Science*, **262**, 1274–1277.

Kaneto, H., Fujii, J., Seo, H.G., Suzuki, K., Matsuoka, T., Nakamura, M., Tatsumi, H., Yamasaki, Y., Kamada, T. and Taniguchi, N. (1995) Apoptotic cell death triggered by nitric-oxide in pancreatic beta cells. *Diabetes*, **44**, 733–738.

Kantrow, S.P. and Piantadosi, C.A. (1997) Release of cytochrome c from liver mitochondria during permeability transition. *Biochem. Biophys. Res. Commun.*, **232**, 669–671.

Kass, G.E.N. (1997) Free-radical-induced changes in cell signal transduction. In: *Free Radical Toxicology*, K.B. Wallace, ed. Taylor & Francis, New York, pp. 349–373.

Kass, G.E.N., Eriksson, J.E., Weis, M., Orrenius, S. and Chow, S.C. (1996) Chromatin condensation during apoptosis requires ATP. *Biochem. J.*, **318**, 749–752.

Kass, G.E.N. and Jones, R.A. (1999) Methods for assessing apoptosis. In: *Approaches to High Throughput Screening*, C.K. Atterwill *et al.*, eds. Taylor & Francis, London, pp. 107–138.

Kass, G.E.N., Juedes, M.J. and Orrenius, S. (1992a) Cyclosporin A protects hepatocytes against prooxidant-induced cell killing. A study on the role of mitochondrial Ca^{2+} cycling in cytotoxicity. *Biochem. Pharmacol.*, **44**, 1995–2003.

Kass, G.E.N., Nicotera, P. and Orrenius, S. (1992b) Calcium-modulated cellular effects of oxidants. In: *Biological Oxidants: Generation and Injurious Consequences*, G.C. Cochrane and M.A. Gimbrone, Jr, eds. Academic Press, San Diego, pp. 133–156.

Kass, G.E.N. and Orrenius, S. (1999) Calcium signaling and cytotoxicity. *Envir. Health Persp.*, **107** (Suppl 1), 25–35.

Kerr, J.F.R., Searle, J., Harmon, B.V. and Bishop, C.J. (1987) Apoptosis. In: *Perspectives on Mammalian Cell Death*, C.S. Potten, ed. Oxford University Press, Oxford, pp. 93–128.

Kerr, J.F.R., Wyllie, A.H. and Currie, A.R. (1972) Apoptosis: a basic biological phenomenon with wide-ranging implications in tissue kinetics. *Br. J. Cancer*, **26**, 239–257.

Kluck, R.M., Bossy-Wetzel, E., Green, D.R. and Newmeyer, D.D. (1997) The release of cytochrome c from mitochondria: a primary site for Bcl-2 regulation of apoptosis. *Science*, **275**, 1132–1136.

Kolesnick, R.N. and Kronke, M. (1998) Regulation of ceramide production and apoptosis. *Annu. Rev. Physiol.*, **60**, 643–665.

Krippner, A., Matsuno-Yagi, A., Gottlieb, R.A. and Babior, B.M. (1996) Loss of function of cytochrome c in Jurkat cells undergoing Fas-mediated apoptosis. *J. Biol. Chem.*, **271**, 21629–21636.

Leist, M., Single, B., Castoldi, A.F., Kuhnle, S. and Nicotera, P. (1997) Intracellular adenosine triphosphate (ATP) concentration: A switch in the decision between apoptosis and necrosis. *J. Exp. Med.*, **185**, 1481–1486.

Leist, M., Single, B., Naumann, H., Fava, E., Simon, B., Kuhnle, S. and Nicotera, P. (1999) Nitric oxide inhibits execution of apoptosis at two distinct ATP-dependent steps upstream and downstream of mitochondrial cytochrome c release. *Biochem. Biophys. Res. Commun.*, **258**, 215–221.

Lemasters, J.J., Nieminen, A.L., Qian, T., Trost, L.C. and Herman, B. (1997) The mitochondrial permeability transition in toxic, hypoxic and reperfusion injury. *Mol. Cell. Biochem.*, **174**, 159–165.

Lennon, S.V., Martin, S.J. and Cotter, T.G. (1991) Dose-dependent induction of apoptosis in human tumour cell lines by widely diverging stimuli. *Cell Prolif.*, **24**, 203–214.

Li, H., Zhu, H., Xu, C.-J. and Yuan, J. (1998) Cleavage of BID by caspase 8 mediates the mitochondrial damage in the Fas pathway of apoptosis. *Cell*, **94**, 491–501.

Li, J.R., Billiar, T.R., Talanian, R.V. and Kim, Y.M. (1997) Nitric oxide reversibly inhibits seven members of the caspase family via S-nitrosylation. *Biochem. Biophys. Res. Commun.*, **240**, 419–424.

Li, P., Nijhawan, D., Budihardjo, I., Srinivasula, S.M., Ahmad, M., Alnemri, E.S. and Wang, X.D. (1997) Cytochrome c and dATP-dependent formation of Apaf-1/caspase-9 complex initiates an apoptotic protease cascade. *Cell*, **91**, 479–489.

Lin, K.T., Xue, J.Y., Nomen, M., Spur, B. and Wong, P.Y.K. (1995) Peroxynitrite-induced apoptosis in HL-60 cells. *J. Biol. Chem.*, **270**, 16487–16490.

Lipton, S.A. and Nicotera, P. (1998) Excitotoxicity, free radicals, necrosis, and apoptosis. *Neuroscientist*, **4**, 345–352.

Liu, X.S., Kim, C.N., Yang, J., Jemmerson, R. and Wang. X.D. (1996) Induction of apoptotic program in cell free extracts – requirement for dATP and cytochrome *c*. *Cell*, **86**, 147–157.

Liu, X.S., Zou, H., Slaughter, C. and Wang, X.D. (1997) DFF, a heterodimeric protein that functions downstream of caspase-3 to trigger DNA fragmentation during apoptosis. *Cell*, **89**, 175–184.

Lizard, G., Gueldry, S., Sordet, O., Monier, S., Athias, A., Miguet, C., Bessede, G., Lemaire, S., Solary, E. and Gambert, P. (1998) Glutathione is implied in the control of 7-ketocholesterol-induced apoptosis, which is associated with radical oxygen species production. *FASEB J.*, **12**, 1651–1663.

Lockshin, R.A. and Williams, C.M. (1965) Programmed cell death. I. Cytology of degeneration in the intersegmented muscles of the Pernyi silkmoth. *J. Int. Physiol.*, **11**, 123–133.

Luo, X., Budihardjo, I., Zou, H., Slaughter, C. and Wang, X.D. (1998) Bid, a Bcl-2 interacting protein, mediates cytochrome c release from mitochondria in response to activation of cell surface death receptors. *Cell*, **94**, 481–490.

Mabley, J.G., Belin, V., John, N. and Green, I.C. (1997) Insulin-like growth factor T reverses interleukin-1 β inhibition of insulin secretion, induction of nitric oxide synthase and cytokine-mediated apoptosis in rat islets of Langerhans. *FEBS Lett.*, **417**, 235–238.

Majno, G. and Joris, I. (1995) Apoptosis, oncosis, and necrosis. An overview of cell death. *Am. J. Pathol.*, **146**, 3–15.

Marini, M., Musiani, D., Sestili, P. and Cantoni, O. (1996) Apoptosis of human lymphocytes in the absence or presence of internucleosomal DNA cleavage. *Biochem. Biophys. Res. Commun.*, **229**, 910–915.

Marzo, I., Brenner, C., Zamzami, N., Jurgensmeier, J.M., Susin, S.A. and Vieira, H.A., Prevost, M.C., Xie, Z.H., Matsuyama, S., Reed, J.C. and Kroemer, G. (1998) Bax and adenine nucleotide translocator cooperate in the mitochondrial control of apoptosis. *Science*, **281**, 2027–2031.

Matsuda, M., Masutani, H., Nakamura, H., Miyajima, S., Yamauchi, A., Yonehara, S., Uchida, A., Irimajiri, K., Horiuchi, A. and Yodoi, J. (1991) Protective activity of adult T-cell leukemia-derived factor (ADF) against tumor necrosis factor-dependent cytotoxicity on U937 cells. *J. Immunol.*, **147**, 3837–3841.

Matsuyama, S., Xu, Q.L., Velours, J. and Reed, J.C. (1998) The mitochondrial F_0F_1-ATPase proton pump is required for function of the proapoptotic protein Bax in yeast and mammalian cells. *Mol. Cell*, **1**, 327–336.

Mayer, M. and Noble, M. (1994) N-acetyl-L-cysteine is a pluripotent protector against cell death and enhancer of trophic factor-mediated cell survival in vitro. *Proc. Natl. Acad. Sci. US*, **91**, 7496–7500.

Melino, G., Bernassola, F., Knight, R.A., Corasaniti, M.T., Nistico, G. and Finazzi-Agro, A. (1997) S-nitrosylation regulates apoptosis. *Nature*, **388**, 432–433.

Messmer, U.K. and Brüne, B. (1996) Nitric oxide (NO) in apoptotic versus necrotic RAW 264. 7 macrophage cell death: The role of NO-donor exposure, NAD$^+$ content, and p53 accumulation. *Arch. Biochem. Biophys.*, **327**, 1–10.

Messmer, U.K., Lapetina, E.G. and Brüne, B. (1995) Nitric oxide-induced apoptosis in RAW-264.7 macrophages is antagonized by protein kinase C-activating and protein kinase A-activating compounds. *Mol. Pharmacol.*, **47**, 757–765.

Mohr, S., Zech, B., Lapetina, E.G. and Brüne, B. (1997) Inhibition of caspase-3 by S-nitrosation and oxidation caused by nitric oxide. *Biochem. Biophys. Res. Commun.*, **238**, 387–391.

Montagna, W. and Wilson, J.W. (1955) A cytologic study of the intestinal epithelium of the mouse after total-body X-irradiation. *J. Nat. Cancer Inst.*, **15**, 1703–1736.

Nicholson, D.W. and Thornberry, N.A. (1997) Caspases: killer proteases. *Trends Biochem. Sci.*, **22**, 299–306.

Nicotera, P., Hartzell, P., Davis, G. and Orrenius, S. (1986) The formation of plasma membrane blebs in hepatocytes exposed to agents that increase cytosolic Ca^{2+} is mediated by the activation of a non-lysosomal proteolytic system. *FEBS Lett.*, **209**, 139–144.

Nobel, C.S.I., Burgess, D.H., Zhivotovsky, B., Burkitt, M.J., Orrenius, S. and Slater, A.F.G. (1997a) Mechanism of dithiocarbamate inhibition of apoptosis: Thiol oxidation by dithiocarbamate disulfides directly inhibits processing of the caspase-3 proenzyme. *Chem. Res. Toxicol.*, **10**, 636–643.

Nobel, C.S.I., Kimland, M., Nicholson, D.W., Orrenius, S. and Slater, A.F.G. (1997b) Disulfiram is a potent inhibitor of proteases of the caspase family. *Chem. Res. Toxicol.*, **10**, 1319–1324.

Nunez, G., Benedict, M.A., Hu, Y.M. and Inohara, N. (1998) Caspases: the proteases of the apoptotic pathway. *Oncogene*, **17**, 3237–3245.

Oberhammer, F., Wilson, J.W., Dive, C., Morris, I.D., Hickman, J.A., Wakeling, A.E., Walker, P.R. and Sikorska, M. (1993) Apoptotic death in epithelial cells: cleavage of DNA to 300 and/or 50 kb fragments prior to or in the absence of internucleosomal fragmentation. *EMBO J.*, **12**, 3679–3684.

Ogasawara, J., Watanabe Fukunaga, R., Adachi, M., Matsuzawa, A., Kasugai, T., Kitamura, Y., Itoh, N., Suda, T. and Nagata, S. (1993) Lethal effect of the anti-Fas antibody in mice. *Nature*, **364**, 806–809.

Orrenius, S., Burkitt, M., Kass, G.E.N., Dypbukt, J.M. and Nicotera, P. (1992) Calcium ions and oxidative cell injury. *Annals Neurol.*, **32**, S33–S42.

Pan, G., Humke, E. and Dixit, V.M. (1998) Activation of caspases triggered by cytochrome c in vitro. *FEBS Lett.*, **426**, 151–154.

Pastorino, J.G., Chen, S.T., Tafani, M., Snyder, J.W. and Farber, J.L. (1998) The overexpression of Bax produces cell death upon induction of the mitochondrial permeability transition. *J. Biol. Chem.*, **273**, 7770–7775.

Peter, M.E. and Krammer, P.H. (1998) Mechanisms of CD95 (APO-1/Fas)-mediated apoptosis. *Curr. Opin. Immunol.*, **10**, 545–551.

Petit, P.X., Goubern, M., Diolez, P., Susin, S.A., Zamzami, N. and Kroemer, G. (1998) Disruption of the outer mitochondrial membrane as a result of large amplitude swelling: the impact of irreversible permeability transition. *FEBS Lett.*, **426**, 111–116.

Raffray, M. and Cohen, G.M. (1997) Apoptosis and neurosis in toxicology: A continuum or distinct modes of cell death? *Pharmacol. Ther.*, **75**, 153–177.

Richter, C. and Kass, G.E.N. (1991) Oxidative stress in mitochondria – its relationship to cellular Ca^{2+} homeostasis, cell-death, proliferation, and differentiation. *Chem.-Biol. Interact.*, **77**, 1–23.

Rudin, C.M. and Thompson, C.B. (1997) Apoptosis and disease: Regulation and clinical relevance of programmed cell death. *Annu. Rev. Med.*, **48**, 267–281.

Saini, K.S. and Walker, N.I. (1998) Biochemical and molecular mechanisms regulating apoptosis. *Mol. Cell. Biochem.*, **178**, 9–25.

Samali, A., Nordgren, H., Zhivotovsky, B., Peterson, E. and Orrenius, S. (1999) A comparison study of apoptosis and necrosis in HepG2 cells: Oxidant-induced

caspase inactivation leads to necrosis. *Biochem. Biophys. Res. Commun.*, **255**, 6–11.

Sarih, M., Souvannavong, V. and Adam, A. (1993) Nitric oxide synthase induces macrophage death by apoptosis. *Biochem. Biophys. Res. Commun.*, **191**, 503–508.

Saunders, J.W. (1966) Death in the embryonic system. *Science*, **154**, 604–612.

Schulze-Osthoff, K., Bakker, A.C., Van Haesebroeck, B., Beyaert, R., Jacob, W.A. and Fiers, W. (1992) Cytotoxic activity of tumor-necrosis-factor is mediated by early damage of mitochondrial functions – evidence for the involvement of mitochondrial radical generation. *J. Biol. Chem.*, **267**, 5317–5323.

Sellins, K.S. and Cohen, J.J. (1987) Gene induction by γ-irradiation leads to DNA fragmentation in lymphocytes. *J. Immunol.*, **139**, 3199–3206.

Servomaa, K. and Rytömaa, T. (1990) UV light and ionizing radiations cause programmed death of rat chloroleukaemia cells by inducing retropositions of a mobile DNA element (L1Rn). *Int. J. Radiat. Biol.*, **57**, 331–343.

Shimaoka, M., Iida, T., Ohara, A., Taenaka, N., Mashimo, T., Honda, T. and Yoshiya, I. (1995) NOC, a nitric oxide-releasing compound, induces dose-dependent apoptosis in macrophages. *Biochem. Biophys. Res. Commun.*, **209**, 519–526.

Shimizu, S., Eguchi, Y., Kamiike, W., Waguri, S., Uchiyama, Y., Matsuda, H. and Tsujimoto, Y. (1996) Retardation of chemical hypoxia-induced necrotic cell-death by Bcl-2 and ICE inhibitors – possible involvement of common mediators in apoptotic and necrotic signal transductions. *Oncogene*, **12**, 2045–2050.

Shimizu, S., Narita, M. and Tsujimoto, Y. (1999) Bcl-2 family proteins regulate the release of apoptogenic cytochrome c by the mitochondrial channel VDAC. *Nature*, **399**, 483–487.

Shoji, Y., Uedono, Y., Ishikura, H., Takeyama, N. and Tanaka, T. (1995) DNA-damage induced by tumor necrosis factor α in L929 cells is mediated by mitochondrial oxygen radical formation. *Immunology*, **84**, 543–548.

Sies, H. (1993) Strategies of antioxidant defense. *Eur. J. Biochem.*, **215**, 213–219.

Slater, A.F.G., Nobel, C.S.I., Maellaro, E., Bustamante, J., Kimland, M. and Orrenius, S. (1995) Nitrone spin traps and a nitroxide antioxidant inhibit a common pathway of thymocyte apoptosis. *Biochem. J.*, **306**, 771–778.

Slee, E.A., Zhu, H.J., Chow, S.C., MacFarlane, M., Nicholson, D.W. and Cohen, G.M. (1996) Benzyloxycarbonyl-Val-Ala-Asp (OMe) fluoromethylketone (Z-VAD. FMK) inhibits apoptosis by blocking the processing of CPP32. *Biochem. J.*, **315**, 21–24.

Strasser, A., Harris, A.W. and Cory, S. (1991) Bcl-2 transgene inhibits T cell death and perturbs thymic self-censorship. *Cell*, **67**, 889–899.

Stridh, H., Kimland, M., Jones, D.P., Orrenius, S. and Hampton, M.B. (1998) Cytochrome *c* release and caspase activation in hydrogen peroxide- and tributyltin-induced apoptosis. *FEBS Lett.*, **429**, 351–355.

Sun, X.-M. and Cohen, G.M. (1994) Mg^{2+}-dependent cleavage of DNA into kilobase pair fragments is responsible for the initial degradation of DNA in apoptosis. *J. Biol. Chem.*, **269**, 14857–14860.

Tada-Oikawa, S., Oikawa, S. and Kawanishi, S. (1998) Role of ultraviolet A-induced oxidative DNA damage apoptosis via loss of mitochondrial membrane potential and caspase-3 activation. *Biochem. Biophys. Res. Commun.*, **247**, 693–696.

Tenneti, L., D'Emilia, D.M. and Lipton, S.A. (1997) Suppression of neuronal apoptosis by S-nitrosylation of caspases. *Neurosci. Lett.*, **236**, 139–142.

Thompson, C.B. (1995) Apoptosis in the pathogenesis and treatment of disease. *Science*, **267**, 1456–1462.

Thornberry, N.A. and Lazebnik, Y. (1998) Caspases: enemies within. *Science*, **281**, 1312–1316.

Trump, B.F. and Berezesky, I.K. (1996) The role of altered [Ca²⁺]ᵢ regulation in apoptosis, oncosis, and necrosis. *Biochim. Biophys. Acta*, **1313**, 173–178.

Trump, B.F., Berezesky, I.K., Chang, S.H. and Phelps, P.C. (1997) The pathways of cell death: oncosis, apoptosis, and necrosis. *Toxicol. Path.*, **25**, 82–88.

Tsujimoto, Y. (1997) Apoptosis and necrosis: intracellular ATP level as a determinant for cell death modes. *Cell Death Diff.*, **4**, 429–434.

Turner, N.A., Xia, F., Azhar, G., Zhang, X.M., Liu, L.X. and Wei, J.Y. (1998) Oxidative stress induces DNA fragmentation and caspase activation via the c-Jun NH2-terminal kinase pathway in H9c2 cardiac muscle cells. *J. Mol. Cell. Cardiol.*, **30**, 1789–1801.

Ueda, N. and Shah, S.V. (1992) Endonuclease-induced DNA damage and cell death in oxidant injury to renal tubular epithelial cells. *J. Clin. Invest.*, **90**, 2593–2597.

Ueda, S., Nakamura, H., Masutani, H., Sasada, T., Yonehara, S., Takabayashi, A., Yamaoka, Y. and Yodoi, J. (1998) Redox regulation of caspase-3(-like) protease activity: Regulatory roles of thioredoxin and cytochrome c. *J. Immunol.*, **161**, 6689–6695.

Uehara, T., Kikuchi, Y. and Nomura, Y. (1999) Caspase activation accompanying cytochrome c release from mitochondria is possibly involved in nitric oxide-induced neuronal apoptosis in SH-SY5Y cells. *J. Neurochem.*, **72**, 196–205.

Ushmorov, A., Ratter, F., Lehmann, V., Droge, W., Schirrmacher, V. and Umansky, V. (1999) Nitric oxide-induced apoptosis in human leukemic lines requires mitochondrial lipid degradation and cytochrome C release. *Blood*, **93**, 2342–2352.

Van den Dobbelsteen, D.J., Nobel, C.S.I., Schlegel, J., Cotgreave, I.A., Orrenius, S. and Slater, A.F.G. (1996) Rapid and specific efflux of reduced glutathione during apoptosis induced by anti-Fas/APO-1 antibody. *J. Biol. Chem.*, **271**, 15420–15427.

Vaux, D.L. and Korsmeyer, S.J. (1999) Cell death in development. *Cell*, **96**, 245–254.

Wallach, D., Kovalenko, A.V., Varfolomeev, E.E. and Boldin, M.P. (1998) Death-inducing functions of ligands of the tumor necrosis factor family: a Sanhedrin verdict. *Curr. Opin. Immunol.*, **10**, 279–288.

Weis, M., Kass, G.E.N. and Orrenius, S. (1994) Further characterization of the events involved in mitochondrial Ca²⁺ release and pore formation by prooxidants. *Biochem. Pharmacol.*, **47**, 2147–2156.

Weis, M., Schlegel, J., Kass, G.E.N., Holmström, T.H., Peters, I., Eriksson, J., Orrenius, S. and Chow, S.C. (1995) Cellular events in Fas/APO-1-mediated apoptosis in JURKAT T lymphocytes. *Exp. Cell Res.*, **219**, 699–708.

Wolfe, J.T., Ross, D. and Cohen, G.M. (1994) A role for metals and free radicals in the induction of apoptosis in thymocytes. *FEBS Lett.*, **352**, 58–62.

Wong, G.H.W., Elwell, J.H., Oberley, L.W. and Goeddel, D.V. (1989) Manganous superoxide dismutase is essential for cellular resistance to cytotoxicity of tumor necrosis factor. *Cell*, **58**, 923–931.

Wyllie, A.H. (1980) Glucocorticoid-induced thymocyte apoptosis is associated with endogenous endonuclease activation. *Nature*, **284**, 555–556.

Yang, J., Liu, X.S., Bhalla, K., Kim, C.N., Ibrado, A.M., Cai, J.Y., Peng, T.I., Jones, D.P. and Wang, X.D. (1997) Prevention of apoptosis by bcl-2: release of cytochrome c from mitochondria blocked. *Science*, **275**, 1129–1132.

Zhivotovsky, B., Cedervall, B., Jiang, S., Nicotera, P. and Orrenius, S. (1994) Involvement of Ca²⁺ in the formation of high molecular weight DNA fragments in thymocyte apoptosis. *Biochem. Biophys. Res. Commun.*, **202**, 120–127.

Zhivotovsky, B., Nicotera, P., Bellomo, G., Hanson, K. and Orrenius, S. (1993) Ca^{2+} and endonuclease activation in radiation-induced lymphoid cell death. *Exp. Cell Res.*, **207**, 163–170.

Zhivotovsky, B., Orrenius, S., Brustugun, O.T. and Døskeland, S.O. (1998) Injected cytochrome c induces apoptosis. *Nature*, **391**, 449–450.

Zoratti, M. and Szabò, I. (1995) The mitochondrial permeability transition. *Biochim. Biophys. Acta*, **1241**, 139–176.

Human and Environmental
Health Effects

6 Dietary Phytoestrogens, Oestrogens and Tamoxifen: Mechanisms of Action in Modulation of Breast Cancer Risk and in Heart Disease Prevention

HELEN WISEMAN

Nutrition, Food and Health Research Centre, King's College London, Franklin-Wilkins Building, 150 Stamford Street, London SE1 8WA, UK

Dietary phytoestrogens, oestrogens and tamoxifen: role in modulation of breast cancer risk and in heart disease prevention

In this chapter the mechanisms by which endogenous oestrogens, dietary phytoestrogens and tamoxifen exert a protective effect against heart disease are critically examined, together with the role of oestrogens (and especially oestrogen metabolism) in increased breast cancer and the protective effects of tamoxifen and potentially phytoestrogens against breast cancer risk.

Oestrogens and risk of breast cancer

Breast cancer is still a major cause of death for women in Western countries. Breast cancer is thought to have a multifactorial causation ranging from gene profile to diet and lifestyle (Wiseman 1994). In addition, mutations in particular tumour suppressor genes such as BRCA1, BRCA2 and p53 are of importance. In addition a functional link of BRCA1 and the ataxia telangiectasia gene product in the DNA damage response has been reported (Li *et al.* 2000) – see Chapter 2. An involvement has been found for the protein p85, a regulator of the signalling protein phosphatidyl-3-OH kinase, in the protein p53-dependent apoptotic response to oxidative stress (Yin *et al.* 1998) – see Chapter 5. Furthermore compromised HOXA5 function has been reported to limit p53 expression in human breast tumours (Raman *et al.* 2000).

The role of endogenous oestrogens in breast cancer is widely recognised. Different forms of oestrogen metabolism result in the formation of mitogenic endogenous oestrogens or the metabolic activation of oestrogens that can result in carcinogenic free-radical mediated DNA damage. Drug and environmental xenoestrogens and metabolites influence breast cancer risk, for example, decreased risk with tamoxifen (see below) and increased risk with oestrogenic pesticides. The likely protective influence of dietary phytoestrogens metabolism on breast cancer risk is discussed below. Pregnancy appears to be important in breast cancer risk: nulliparous women have the greatest risk of breast cancer, but for parous women multipregnancies are of no greater benefit than a single pregnancy (Peeters *et al.* 1995). An earlier pregnancy is more protective than a later one because of the differentiation of breast epithelial cells into milk-producing cells that occurs in the breast during pregnancy (Russo and Russo 1995) and the apoptosis of breast cells that occurs in the breast following pregnancy (chance for elimination of mutated epithelial cells) (Barnes 1998).

Oestrogen receptor in ligand-binding events

Oestrogens are recognised to be involved in the growth, development and homeostasis of oestrogen-responsive tissues. The oestrogen receptor mediates the biological activity of oestrogens and is a ligand-inducible nuclear transcription factor: oestrogen binds to the ligand-binding domain of the oestrogen receptor, triggering either the activation or repression of target genes (Katzenellenbogen *et al.* 1996).

The selective oestrogen receptor antagonist raloxifene (structurally related to tamoxifen) inhibits the mitogenic effects of oestrogen in reproductive tissues, while maintaining the beneficial effects of oestrogen in other tissues. The crystal structures have been reported of the ligand-binding domain of the oestrogen receptor complexed to either 17β-oestradiol or to raloxifene (Brzozowski *et al.* 1997), providing structural evidence for the mechanisms of oestrogen receptor agonism and antagonism (agonist and antagonist bind at the same site within the core of the ligand-binding domain, but with a different binding mode). The activation function-2 (AF-2) helix in the agonist-bound oestrogen receptor structure (also retinoic-acid receptors and thyroid hormone receptors) is folded against the ligand-binding domain and the AF-2 hydrophobic residues form part of the ligand-binding pocket. The switch from agonism to antagonism appears to involve a ligand-mediated repositioning of helix 12 of the ligand binding domain of the oestrogen receptor such that a transcriptionally competent transcriptional AF-2 is no longer generated

(Brzozowski *et al.* 1997). A combination of specific polar and non-polar interactions enables the oestrogen receptor to selectively recognise and bind 17β-oestradiol with subnanomolar affinity, the oestrogen receptor is the only steroid receptor able to additionally interact with a large number of non-steroidal compounds. Phytoestrogens and environmental and drug xenoestrogens (and their metabolites) often show a structural similarity to the steroid nucleus of oestrogen: most particularly a phenolic ring analogous to the A-ring in oestradiol. These structural features enable them to bind to oestrogen receptors to elicit responses ranging from agonism to antagonism of the endogenous hormone ligand (Miksicek 1995).

It was originally accepted that only one oestrogen receptor existed (the classical oestrogen receptor is now termed ERα), in contrast to other members of the nuclear receptor superfamily where multiple forms have been found; however, a separate subtype, termed oestrogen receptor β (ERβ) has been identified in cDNA libraries from human testis and rat prostate tissues (Kuiper *et al.* 1996). ERβ has a different tissue distribution from ERα and is most strongly expressed in ovary, uterus, prostate and lung (Kuiper *et al.* 1997). Expression of ERβ in the breast appears to be low and additionally it is expressed in different sites in the brain to ERα (Barnes 1998). Furthermore, the α and β subtypes of oestrogen receptor display high conservation of amino acid sequence in regions of the ligand-binding domain important in ligand binding (Kuiper *et al.* 1997). This indicates that alterations in ligand structure to ensure preferential selectivity for one type of receptor subtype may be difficult to achieve. Moreover, although some studies on the binding of oestrogens, xenoestrogens and phytoestrogens (see below) have been reported (Ekena *et al.* 1996), studies focusing on the relative binding affinities of the parent compound compared to its metabolites are still required.

Oestrogen metabolism and the control of oestrogen levels

Oestrogen metabolism to form the biologically active oestrogen, oestradiol, is catalysed by steroid sulphatase, oestradiol 17β-hydroxysteroid dehydrogenase and aromatase (Purohit *et al.* 1999). A role for oestrogens in the promotion of breast cancer is likely. Several studies have indicated a direct relationship in post-menopausal women between breast cancer risk and plasma oestradiol concentration (Thomas *et al.* 1997a, b). In the 'Guernsey study', plasma samples from 61 women who went on to develop breast cancer (on average around 7.8 years later) were matched with 179 from women who did not develop breast cancer and the relative risk was found to be five times higher in the upper compared to the lower tertile of plasma oestradiol levels (Thomas *et al.* 1997a; Dowsett 1998). Adjustment for oestradiol removed the twofold relative risk found for testosterone, suggesting that the risk with testosterone derived from it being a substrate for the aromatase enzyme complex which converts it to oestradiol. The aromatase enzyme complex also converts androstenedione to oestrone by the same pathway. The gradient of this risk relationship is greater than expected from data on women on hormone replacement therapy (HRT), perhaps because aromatase activity raises oestrogen levels in the breast tissue itself. Indeed, although aromatase activity in breast carcinoma is very variable, in some instances it may be enough to support breast cancer growth independent of plasma oestradiol levels, and reports suggest that different regulation of aromatase in normal and malignant breast tissue may occur.

Steroid sulphatase and oestradiol 17β-hydroxysteroid dehydrogenase (Type 1) activity contribute to the high concentrations of oestrogens that are found in breast tumours (activities of these enzymes are increased in breast cancer tissue) (Purohit et al. 1999). Steroid sulphatase regulates the formation of unconjugated steroids such as oestrone and dehydroepiandrosterone from their sulphated conjugates, which are unable to interact with steroid receptors. This enzyme has a vital role therefore in regulating tissue steroid availability (Purohit et al. 1999). Oestradiol 17β-hydroxysteroid dehydrogenase (Type 1) converts oestrone and dehydroepiandrosterone to oestradiol and androstenediol (Adiol), respectively. Both oestradiol and androstenediol bind to the oestrogen receptor and stimulate the growth of breast cancer cells in vitro and carcinogen-induced mammary tumours in vivo. A number of factors are important in the stimulation of the activities of these enzymes including the cytokines interleukin-6 and tumour necrosis factor-α (Purohit et al. 1999).

Oestrogen metabolism and catecholoestrogen formation

Conversion of oestradiol to its metabolite 4-hydroxyoestradiol (this extrahepatic oestrogen 4-hydroxylase activity has been identified as cytochrome P-4501B1) predominates over the more usual and non-genotoxic 2-hydroxyoestradiol formation (cytochrome P-450 3A in liver and cytochrome P-450 1A in extra hepatic tissue) in neoplastic breast tissue compared to normal breast tissue (Liehr 1999). An elevated 4-hydroxyoestradiol/2-hydroxyoestradiol ratio may thus be a useful marker of malignant breast tumours (Liehr 1999). In addition, the ratio of genotoxic 16α-hydroxyoestrone to protective 2-hydroxyoestrone (weakly oestrogenic and can inhibit breast cell proliferation), may be a useful marker of breast cancer risk that can be measured, for example, in urine (Pasagian-Macaulay et al. 1996). Some xenoestrogens, including environmental oestrogens (see Chapter 4), for example oestrogenic pesticides such as kepone, can increase the 16α-hydroxyoestrone/2-hydroxyoestrone ratio and are thus likely to increase breast cancer risk (Bradlow *et al.* 1995). Other xenoestrogens such as the antioestrogen drug tamoxifen, used successfully in both the treatment and prevention of breast cancer, and its more potent antioestrogen 4-hydroxy metabolite are protective against breast cancer (Wiseman 1994).

Free radicals are likely to have a direct role in human cancer development as they have been shown to possess many characteristics of chemical carcinogens (Wiseman and Halliwell 1996). Elevated plasma or urine concentrations of oestrogens are a known risk factor for breast cancer in humans. Although oestrogens can induce the proliferation of cells by receptor-initiated mechanisms, there is likely to be a role for metabolic activation of oestrogens in tumour initiation (Liehr 1999). Indeed, an elevated 4-hydroxylation of oestradiol by a specific 4-hydroxylase has been found in human breast and in mammary tumours (Liehr 1999). Moreover, catecholoestrogens including 4-hydroxyoestradiol may undergo metabolic redox cycling between hydroquinone and quinone forms, which is a mechanism of free radical generation and both direct and indirect free radical-mediated DNA damage has been induced by oestrogens both *in vitro* and *in vivo* (Liehr 1999). Free radical-mediated DNA damage including the formation of 8-hydroxyguanine (Malins *et al.* 1993) and DNA adducts formed by malondialdehyde, a decomposition product of free radical-induced lipid peroxides (Wang *et al.* 1996) have been found in the DNA of breast cancer patients (see Chapter 2).

Membrane-antioxidant action in protective mechanisms

Steroid hormones act at the genomic level by binding to nuclear receptors and subsequently modulating the expression of specific target genes (see above); recent research, however, has shown that direct interaction between steroid hormones and G-protein-coupled receptors is also possible (progesterone has been shown to inhibit the oxytocin receptor by direct binding: Grazzini *et al.* 1998). It is of related interest that a mechanism has recently been proposed for the molecular activation of the trimeric G protein on the inside of the cell membrane in response to the hormonal signals transmitted by transmembrane receptors, resulting in the relaying of the signals to intracellular enzymes and ion channels; in this model the membrane-bound receptor uses two complementary mechanisms to act at a distance on the G proteins's guanine-nucleotide-binding pocket (Iiri *et al.* 1998). Moreover, the integrity of membrane function is now known to be vital to many cellular processes, including the role of membrane enzymes and receptors in cell growth, and signalling thus protection of membrane function against the free radical-mediated process of membrane lipid peroxidation that can result in oxidative membrane damage is clearly of great importance (Wiseman 1996; Wiseman 1998). The membrane antioxidant properties of tamoxifen (Wiseman 1994) and soya isoflavones (Wiseman *et al.* 1998) have been suggested to contribute to their ability to decrease breast cancer risk, see below. In model membrane systems equol (daidzein metabolite) was a more potent antioxidant than genistein or daidzein itself, similarly the tamoxifen metabolite, 4-hydroxytamoxifen was more potent. Tamoxifen and 4-hydroxytamoxifen appear to act as membrane antioxidants, at least in part by entering the membrane and modulating membrane fluidity which may thus increase membrane stability against lipid peroxidation (Wiseman *et al.* 1993) and a similar mechanism may be displayed by isoflavones. Overall this suggests an important role for metabolism in increasing the antioxidant effectiveness of this drug and the isoflavone phytoestrogens. It should be noted, however, that 17β-oestradiol is also a good membrane antioxidant (Wiseman *et al.* 1990), and although this is important in relation to its beneficial cardioprotective actions, it does not appear to prevent it from being a major risk factor for breast cancer, although on balance this may be true for some individuals.

Oestrogens in cardioprotection

The role of oestrogens in cardioprotection is currently the focus of considerable attention. Oestrogen administration in post-menopausal women has been observed to produce cardioprotective benefits (Stampfer *et al.* 1991). However, the exact biomolecular mechanisms for this cardioprotection remain unclear. It is likely that actions mediated both through the oestrogen receptor (genomic effects), such as the beneficial alteration in lipid profiles and upregulation of the low-density lipoprotein (LDL) receptor, and independently of the oestrogen receptor (non-genomic effects), such as antioxidant action, contribute to the observed cardiopro-tective effects of oestrogens.

Mechanisms of cardioprotection by oestrogen and oestrogen mimics

Antioxidant properties *in vitro* have been shown for the oestrogens, 17β-oestradiol, 17α-ethynyloestradiol, genistein, tamoxifen (in particular 4-hydroxytamoxifen) and phloretin (in particular 3-hydroxyphloretin, see Chapter 1). 17β-Oestradiol (Sack *et al.* 1994) and tamoxifen (Guetta *et al.* 1995) have also been shown to inhibit LDL oxidation following their administration to post-menopausal women. This is important because oxidative damage to LDL has been implicated in atherogenesis (see Chapter 1). Oestrogen may be able to enhance beneficial relaxation of the vasculature by influencing the synthesis of vasodilators such as the free radical nitric oxide (Ruehlmann and Mann 1997). In addition, oestrogens can act rapidly on smooth muscle calcium signalling (Ruehlmann and Mann 1997) and a possible modulation of membrane protein conformation (particularly of receptors, channels or enzymes), resulting from changes in membrane fluidity, may contribute to some of these effects. Oestrogens are known modulators of membrane fluidity (Wiseman 1994): they mimic the membrane stabilisation against lipid peroxidation demonstrated by the natural membrane fluidity modulator cholesterol (Wiseman 1994, see Chapter 1). The membrane antioxidant properties of oestrogens may also contribute to their ability to protect against muscle damage (Bar and Amelink 1997) including, potentially, damage to the heart muscle. Another important beneficial effect of oestrogens is the lowering of plasma homocysteine levels (Lien et al. 1997). This is because elevated levels of this amino acid is a known risk factor for cardiovascular disease. In addition to the favourable influence of oestrogen on plasma lipid profiles, oestrogens improve lipid metabolism in the postprandial state. In particular, improvements in the clearance of chylomicron remnants and attenuation of the postprandial decrease in levels of beneficial high-density lipoprotein (HDL)-cholesterol in the presence of oestrogens, make this state less atherogenic (Westerveld et al. 1997).

Future prospects for cardioprotection by oestrogen and oestrogen mimics

The production and use of novel dietary oestrogens with improved cardidoprotective properties is a possibility (Ridgway and Tucker 1997). It could also be advantageous to use structural modification to remove the oestrogenicity of novel compounds. This would enable their use in a wider range of products (Ridgway and Tucker 1997). New compounds such as 3-hydroxyphloretin (a derivative of the oestrogenic aglycone form of phloridzin, see Chapter 1), which combine greatly enhanced antioxidant activity with decreased oestrogenicity, could thus be of potential use. However, the beneficial oestrogen receptor-mediated cardioprotective effects of these compounds would also be lost. Thus the production of cocktails of compounds with optimised oestrogenic and antioxidant properties, as appropriate for different cardioprotective requirements, is to be expected soon (Ridgway and Tucker 1997).

The anticancer drug tamoxifen

Tamoxifen is an anticancer drug, now known not only for its role in the treatment of breast cancer, but also for the controversy surrounding its use as a prophylactic agent in the prevention of breast cancer in healthy women who are considered from their family history to be at risk of developing this disease (Wiseman 1994; Kellen 1996; Jordan 1997). Tamoxifen is a tumoristatic (rather than tumoricidal) drug, suggesting its use as a prophylactic, primary and/or secondary prevention drug for breast cancer. It has been suggested that tamoxifen would either halt the onset of cancer or alternatively increase the length of time before the clinical appearance of a breast tumour (Love 1992). A randomised, double-blind clinical trial involving 13 388 women aged over 65 years of age and at high risk of developing breast cancer based on family history and other factors, was sponsored for around four years by the National Cancer Institute, USA. In April 1998, the trial was closed because it was found that the women receiving tamoxifen had a 45% lower incidence of breast cancer compared with those on placebo treatment (NCI 1998) and it was considered unethical to continue with the placebo treatment in women who might benefit from an effective therapy. There are, however, several disadvantages with long-term prophylactic use of tamoxifen including an increased risk of endometrial cancer, pulmonary embolism and deep vein thrombosis (these are potential problems with all oestrogenic therapies including HRT and the contraceptive pill). The potential of other oestrogenic substances to prevent breast cancer is currently being investigated and candidates include the synthetic oestrogens such as raloxifene (strongly indicated as an HRT may be particularly protective against bone loss (Mitlak and Cohen 1997)): the first of a class of selective oestrogen receptor modulators (SERMs). Tamoxifen itself is an emerging SERM with potential in HRT under certain conditions (Mitlak and Cohen 1997). Dietary phytoestrogens (soya isoflavones and lignans, particularly from linseed) may also be of value and are likely to be less toxic than synthetic oestrogens, although care is needed in the amounts that could be self-administered from available food supplements.

In addition to protective effects against breast cancer, substantial cardioprotective effects of tamoxifen are clearly emerging. This is very important because although hormonally related cancers such as breast cancer are the most common cause of mortality in women aged 40–60 years old, cardiovascular disease is the leading cause of death in women over the age of 60. Developments in this area include protection by tamoxifen of LDL particles against the oxidation implicated in atherosclerosis (Wiseman *et al* 1993; Guetta *et al.* 1995) lowering by tamoxifen of both homocysteine levels (Anker *et al.* 1995) and lipoprotein(a) levels (Shewmon *et al.* 1994a, b) which are both risk factors for atherosclerosis and elevation by tamoxifen of secretion of the protective transforming growth factor-β (Grainger *et al.* 1995a, b). In addition, tamoxifen has direct effects on arterial LDL metabolism: in monkeys tamoxifen has been shown to inhibit arterial accumulation of LDL degradation products and progression of coronary artery atherosclerosis (Williams *et al.* 1997).

Tamoxifen metabolism in humans

The absorption, distribution, metabolism and excretion of tamoxifen has been extensively reviewed (Furr and Jordan 1984; Buckley and Goa 1989; Wiseman 1994, 1996). Tamoxifen is well absorbed after oral administration and appears to be greater than 99% bound to plasma proteins (mostly to albumin) (Lien *et al.* 1989). However, results from studies of tamoxifen absorption in volunteers show large inter-individual variation and this is likely to be due to differences in liver metabolism and differences in absorption in the gastrointestinal tract (McVie *et al.* 1986). Administration of 40 mg/day for two months to patients with breast cancer produced steady-state plasma concentrations of tamoxifen of 186–214 ng/ml (McVie *et al.* 1986). Administration of a single 20 mg dose of tamoxifen to male volunteers resulted in a peak plasma concentration of tamoxifen of 42 ng/ml and N-desmethyltamoxifen of 12 ng/ml (Adam *et al.* 1980). Steady-state concentrations of tamoxifen were reached after three to four weeks' administration and of N-desmethyltamoxifen after eight weeks (Adam *et al.* 1980). Administration of 40 mg/day of tamoxifen (Daniel 1981) resulted in plasma concentrations of tamoxifen of 27–520 ng/ml (mean 300) and tumour biopsy concentrations of tamoxifen of 5.4–117 ng/mg (mean 25.1) protein; plasma concentrations of N-desmethyltamoxifen of 210–761 ng/ml (mean 462) and tumour biopsy concentrations of N-desmethyltamoxifen of 7.8–210 ng/mg (mean 52) protein and plasma concentrations of 4-hydroxytamoxifen of 2.8–11.4 ng/ml (mean 6.7) and tumour biopsy concentrations of 4-hydroxytamoxifen of 0.29–1.13 ng/mg (mean 0.53) protein.

At this daily dose of tamoxifen (40 mg) after 14 days plasma tamoxifen concentrations in three patients (Murphy *et al.* 1987) were 363 ng/ml (8.04 and 8.14 ng/mg protein in breast tumour nuclear fraction and cytosol respectively), 745 ng/ml (11.09 and 18.52 ng/mg protein in breast tumour nuclear fraction and cytosol respectively) and 406 ng/ml (9.36 and 17.41 ng/mg protein in breast tumour nuclear fraction and cytosol respectively). In these same three patients N-desmethyltamoxifen plasma concentrations were 185 ng/ml (3.56 and 6.60 ng/mg protein in breast tumour nuclear fraction and cytosol respectively), 311 ng/ml (7.01 and 26.75 ng/mg protein in breast tumour nuclear fraction and cytosol respectively), 422 ng/ml (7.87 and 14.10 ng/mg protein in breast tumour nuclear fraction and cytosol respectively). For 4-hydroxytamoxifen plasma concentrations were 2.5 ng/ml (0.159 and 0.022 ng/mg protein in breast tumour nuclear fraction and cytosol respectively), 3.1 ng/ml (0.226 and 0.275 ng/mg protein in breast tumour nuclear fraction and cytosol respectively), 1.4 ng/ml (0.258 and 0.357 ng/mg protein in breast tumour nuclear fraction and cytosol respectively).

Studies have shown that in humans, levels of tamoxifen and its metabolites were 10-fold to 60-fold higher in tissues than in serum and relatively high concentrations were detected in liver and lung (Lien *et al.* 1991). Pancreas, pancreatic tumour and brain metastases from breast cancer and primary breast cancer also retained large amounts of the drug. The amounts of N-demethylated and hydroxylated metabolites were high in most tissues except in fat, and tamoxifen and some metabolites were also present in specimens of skin and bone tissue (Lien *et al.* 1991). The distribution half-life of tamoxifen is 7–14 hours (Fromson *et al.* 1973b). A study using 10 mg of tamoxifen showed that the plasma elimination half-life of tamoxifen of approximately 20 hours, determined over the first 24 hour values, was a distribution half-life (de Vos *et al.* 1989).

Tamoxifen is extensively metabolised by cytochrome P-450-dependent hepatic mixed function oxidases to form tamoxifen metabolites including N-desmethyltamoxifen (the major metabolite) and 4-hydroxytamoxifen (more active than tamoxifen, but only present in very small amounts). In addition, 4-hydroxy-N-desmethyltamoxifen has been identified as another tamoxifen metabolite in human bile (Lien *et al.* 1988) and has also been identified in plasma (Lien *et al.* 1989). The cytochrome P-450 CYP3A and CYP1A1 enzymes are thought to be involved in the *n*-demethylation of tamoxifen in human liver microsomes (Jacolot *et al.* 1991; Mani *et al.* 1993, 1994). In healthy male volunteers given 40 mg of tamoxifen, the tamoxifen plasma elimination half-life during day 1 was 10 hours. However, after 34 hours appreciable levels of tamoxifen and N-desmethyltamoxifen were still present, suggesting a lengthening half-life with increasing study duration (Guelen *et al.* 1987). Elimination of tamoxifen appears to be bi-phasic with an initial phase of 7–14 hours in female patients and a terminal phase of around seven days (Fromson *et al.* 1973b). The elimination of N-desmethyltamoxifen is around seven days and the 4-hydroxtamoxifen has a shorter half-life than tamoxifen (Buckley and Goa 1989). Tamoxifen and its metabolites are mostly excreted via bile into faeces as glucuronides and other conjugates. Urinary excretion is only a very minor route of elimination of the drug (Furr and Jordan 1984).

Tamoxifen as a protective agent against heart disease

Influence of tamoxifen on cardiovascular risk factors

Retrospective studies on the two randomised arms of the Scottish adjuvant tamoxifen trial has revealed a significant decrease in the incidence of fatal myocardial infarction in breast cancer patients treated with tamoxifen (McDonald and Stewart 1991). Tamoxifen therapy for at least five years appears to have cardioprotective action in post-menopausal women similar to that of oestrogen. The Stockholm randomised trial of adjuvant tamoxifen therapy in post-menopausal women with early-stage breast cancer reported that even short-term tamoxifen therapy significantly decreased occurrence of coronary heart disease (Rutqvist and Mattsson 1993).

Treatment with tamoxifen as an adjuvant in a group of 123 women with stage I and II breast cancer produced a significant decrease in serum cholesterol levels compared with a control group of 81 women with stage I and II breast cancer who were not taking a hormonal treatment or supplement (Schapira *et al.* 1990). In this study decreases in serum cholesterol of more than 10 mg/dl were recorded in 73% of tamoxifen-treated patients compared to 35% of controls. Decreases in cholesterol of more than 40 mg/dl were found in 40% of tamoxifen-treated patients compared to 13% of controls.

In the Wisconsin Tamoxifen Study, a randomised placebo-controlled double-blind study in disease-free women, tamoxifen treatment was found to lower total cholesterol levels by 26 mg/dl and LDL cholesterol levels by 20% after three months of treatment (Love *et al.* 1991). These changes were maintained throughout the two years of tamoxifen treatment, and women with greater baseline cholesterol levels showed greater decreases with tamoxifen treatment. The Wisconsin Tamoxifen Study also reported that fibrinogen levels decreased by 52 mg/dl after six months and this may be associated with a decreased risk of arterial thrombosis (Love *et al.* 1992). In a two-year, randomised, placebo-controlled trial of 20 mg/day tamoxifen in 57 normal post-menopausal New Zealand women, tamoxifen treatment lowered levels of serum cholesterol by a mean of $12 \pm 2\%$, LDL by $19 \pm 3\%$ and fibrinogen by $18 \pm 4\%$ (Grey *et al.* 1995). Furthermore, the favourable changes in lipid, lipoprotein and fibrinogen levels seen in early tamoxifen therapy persist with treatment of five years (Love *et al.* 1994).

Protection by tamoxifen against LDL oxidation

Tamoxifen, and in particular 4-hydroxytamoxifen, have been reported to protect isolated human LDL against oxidative modification (Wiseman *et al.* 1993). Clinical evidence has now been obtained for the ability of tamoxifen to achieve this protection when administered to patients (Guetta *et al.* 1995). Protection of LDL against oxidative damage by tamoxifen is likely to be of considerable importance because oxidative damage to LDL is a critical stage in the development of atherosclerosis. It is a prerequisite for macrophage uptake and cellular accumulation of cholesterol (Steinberg *et al.* 1989; Witzum 1994; Steinberg and Lewis 1997). In a study on the action of tamoxifen and related compounds on oxidative damage to isolated human LDL, 4-hydroxytamoxifen was more effective as an inhibitor of Cu(II) ion-dependent lipid peroxidation than tamoxifen and also prevented peroxidation-induced modifications in the surface charge of the LDL, whereas tamoxifen did not (Wiseman *et al.* 1993). These alterations in the surface charge of LDL are associated with its recognition and uptake by macrophages in atherosclerotic lesions. This action of the 4-hydroxy metabolite of tamoxifen could be particularly important to the observed beneficial cardiovascular effects of tamoxifen therapy. The lack of effectiveness of tamoxifen itself in preventing alteration of the surface charge of LDL may be because it is a much less effective inhibitor of lipid peroxidation than 4-hydroxytamoxifen. Tamoxifen, over the concentration range tested (5–30 μM), maximally inhibited LDL peroxidation by only approximately 50% compared to approximately 90% achieved by 4-hydroxytamoxifen. In the presence of tamoxifen, therefore, it is likely that sufficient aldehydic breakdown products of lipid peroxidation such as malondialdehyde and 4-hydroxynonenal are produced to modify the ε-amino groups of the lysine residues of the apolipoprotein B molecule and thus the surface charge of LDL (Wiseman *et al.* 1993). Tamoxifen and its derivatives are highly lipophilic compounds that are likely to accumulate in the atheromal plaques associated with the damaged arterial wall to achieve the protective concentrations reported.

Tamoxifen administered for two months to post-menopausal women significantly reduced plasma LDL levels and protected LDL from oxidation (Guette *et al.* 1995). This was measured by the time of the onset of LDL oxidation (i.e. the lag time in the presence of Cu(II) ions) in LDL isolated from women treated with tamoxifen. The lag time to onset of LDL oxidation was $20 \pm 35\%$ longer during the tamoxifen treatment period than during treatment with placebo. This decreased susceptibility of the LDL to oxidation persisted only as long as the tamoxifen treatment. Tamoxifen (and 4-hydroxytamoxifen) may stabilise LDL against lipid peroxidation by interactions between their hydrophobic rings and the polyunsaturated residues of the phospholipid layers of LDL. This suggestion is supported by the inhibition of lipid peroxidation arising from similar interactions in liposomal membranes (Wiseman *et al.* 1990b).

It would be useful to determine whether tamoxifen can also protect against the reported homocysteine-induced oxidation of LDL (Hirano *et al.* 1994), in addition to its ability to lower plasma homocysteine levels (Anker *et al.* 1995). The ability of tamoxifen to protect against LDL oxidation in women is further important evidence in favour of the cardioprotective benefits of long-term tamoxifen therapy.

Ability of tamoxifen to decrease atherogenic homocysteine levels

Tamoxifen has been reported to decrease plasma levels of the amino acid homocysteine (Anker *et al.* 1995). This may also contribute to its cardioprotective action because elevated plasma levels of homocysteine is a known risk factor for cardiovascular disease. Even moderately elevated homocysteine levels have been reported to be associated with an increased risk of premature cardiovascular disease. In a study of 31 post-menopausal women with breast cancer the plasma homocysteine level was decreased by a mean value of 30% after 9–12 months of tamoxifen treatment. Elevated homocysteine levels are associated with premature vascular disease and thrombosis, and it is of interest that homocysteine can induce tissue factor procoagulant activity in endothelial cells (Fryer *et al.* 1993). In addition, homocysteine can induce the oxidative damage to LDL implicated in atherosclerosis (Hirano *et al.* 1994). Oestrogen status may be an influencing factor on homocysteine levels: pregnancy and oestrogen replacement therapy decrease plasma homocysteine levels (Andersson *et al.* 1992), and tamoxifen may be acting like oestrogen to produce this beneficial effect on plasma homocysteine levels.

Ability of tamoxifen to decrease atherogenic lipoprotein(a) levels

Tamoxifen has been reported to decrease the circulating levels of lipoprotein(a) (Lp(a)) in healthy post-menopausal women and in breast cancer patients (Shewmon *et al.* 1994a, b). In healthy post-menopausal women plasma Lp(a) levels decreased to 24% below baseline after one month of treatment with tamoxifen, and after three months this further decreased to 34% below baseline (Shewmon *et al.* 1994b). In addition, insulin like growth factor (IGF-1) levels decreased to 30% below baseline after three months treatment with tamoxifen and the correlation between Lp(a) and IGF-1 was highly significant, which may indicate that they share regulatory influences (Shewmon *et al.* 1994b). Furthermore, tamoxifen lowered Lp(a) levels by 40% after only one month of therapy in patients with breast cancer (Shewmon *et al.* 1994a).

This lowering of Lp(a) levels by tamoxifen could be of particular importance because lipoprotein(a) levels may account for much of the previously unattributable risk for coronary heart disease (Hayden and Reidy 1995). Tamoxifen is therefore an alternative to niacin for the treatment of elevated Lp(a) levels. Lipoprotein(a) differs structurally from LDL only by an extra protein molecule, apolipoprotein(a), which resembles plasminogen and is linked to apolipoprotein B-100. The reasons for the extreme atherogenicity of lipoprotein(a) are not yet clear, but following oxidative modification it may promote macrophage transformation to foam cells (even more effectively than oxidised LDL) and it may directly promote the growth of atherosclerotic plaque. Lipoprotein(a) is also thought to prevent the conversion of plasminogen to the antithrombotic plasmin.

Elevation by tamoxifen of protective TGF-β levels

The cytokine transforming growth factor-β (TGF-β) is thought to play a key role in maintaining the normal structure and function of the cells of the artery wall. Important clinical evidence for the cardioprotective effect of TGF-β is the finding that the plasma levels of active TGF-β are depressed in patients with severe atherosclerosis (Grainger and Metcalf 1996). It may therefore be useful to elevate TGF-β activity in the artery wall. Tamoxifen has been shown to be a potent stimulator of TGF-β in vascular smooth muscle cells, which is consistent with the ability of tamoxifen to elevate TGF-β in breast tumours *in vivo*.

Tamoxifen can induce TGF-β in the aorta and circulation in C57B16 mice and on the influence of tamoxifen on the development of the high-fat diet induced lipid lesions and fatty streak lesions (these resemble the early stages of atherosclerosis in humans), which can be induced in these mice (Grainger et al. 1995). Tamoxifen was found to strongly inhibit the formation of the lipid lesions induced by a high-fat diet in this susceptible strain of mice (Grainger et al. 1995). Furthermore, tamoxifen elevated TGF-β in the aortic wall and in the circulation, which prevents vascular smooth muscle cell activation and thus inhibits lipid accumulation in mice fed a high fat diet. In addition, to inducing the beneficial effects of TGF-β, it is likely that tamoxifen may be acting to modify the structure of the vessel wall itself to prevent the changes in vessel wall structure that make the cells more susceptible to taking up lipid. This could also be via changes to the membranes of the vessel cells: tamoxifen is known to modify membrane fluidity thus making the cells more rigid (Wiseman et al. 1993b) which could inhibit the action of membrane-mediated lipid uptake mechanism. This is more probable than modification of the lipid itself (i.e. protection of LDL against oxidation), which is likely to be a major protective mechanism of tamoxifen in humans. This is because mice (especially on a low fat diet) do not have much circulating LDL, thus prevention of LDL oxidation is unlikely to be the main protective mechanism.

Future directions for tamoxifen as a cardioprotectant?

There is strong indication therefore that tamoxifen has a wide range of complementary beneficial effects on the risk factors that contribute to cardiovascular disease. The rapid progress being made in this exciting field will undoubtedly influence the future direction for the use of tamoxifen for the prevention of breast cancer. In addition, the breakthroughs in determining the genes that may dispose individual women to breast cancer (Davies 1995) will in the future be invaluable in selecting suitable candidates for prophylaxis with tamoxifen, while helping to prevent inappropriate exposure in women for whom the risk is not considered sufficient for this course of action.

However, as the cardioprotective benefits of tamoxifen treatment become better understood and appropriate screening measures are routinely made to detect any undesirable endometrial changes, we may well see a situation where women actively chose tamoxifen prophylaxis because of a family history of cardiovascular disease. A further possibility is whether men with a family history of cardiovascular disease may also actively seek tamoxifen prophylaxis. In support of this is the current use of tamoxifen to treat other cancers in addition to breast cancer. Although a tamoxifen derivative tamandron has been developed for the treatment of prostatic cancer, tamoxifen itself may also offer some protection in a similar way to that claimed for phytoestrogens such as genistein and daidzein found in soya products and reported to be protective against prostatic cancer in Japanese men (see below).

Dietary phytoestrogens

Dietary phytoestrogens are being investigated to assess their importance in the prevention of major diseases, including heart disease and cancer. Phytoestrogens such as the isoflavones genistein and daidzein are the subject of many studies (see reviews by Adlercreutz et al. 1995; Adlercreutz 1996; Cassidy 1996; Bingham et al. 1998; Barnes 1998; Wiseman 1998, 1999). Some members of the flavonoid family including the isoflavonoids and some of the flavones and flavonones, also belong to the phytoestrogen family (non-steroidal oestrogen mimics derived from plants), which also includes some chalcones, coumestans, stilbenes and lignans (Wiseman 1998). Most flavonoids are present in plants in their glycoside form, whereas the unconjugated (aglycone) forms demonstrate greater oestrogenic activity (Miksicek 1995).

Dietary sources, metabolism and bioavailability of isoflavone and lignan phytoestrogens

Dietary sources

The isoflavonoids include the isoflavone phytoestrogens genistein and daidzein, which occur mainly as glycosides of the parent aglycone in soya beans and consequently in a wide range of soya-derived products and to a lesser extent in other legumes (Reinli and Block 1996). Soya foods traditionally made from soya beans, include both fermented and non-fermented foods (Golbitz 1995). In non-fermented soya foods the isoflavones are mostly present as β-glucosides, some of which are esterified with malonic acid or acetic acid (Barnes et al. 1994). In fermented soya foods such as miso or tempeh, the isoflavones are mostly unconjugated (Coward et al. 1993). Some alcoholic beverages such as beer contain significant amounts of isoflavonoid phytoestrogens (Lapcik et al. 1998). Lignans are non-flavonoid dietary phytoestrogens, which are found as the plant precursors (secoisolariciresinol and matairesinol) of the mammalian lignans enterodiol and enterolactone in plant material. Secoisolariciresinol and matairesinol are found in linseed (flaxseed) and sesame seed respectively, and both are also found in a number of whole grains and many other foods (Thompson 1994; Slavin et al. 1997).

Metabolism and bioavailability

Dietary daidzin and genistin are hydrolysed in the large intestine (by the action of bacteria) to release genistein and daidzein. Daidzein can be metabolised by the bacteria in the large intestine to form the isoflavan equol (oestrogenic) or O-desmethylangolensin (non-oestrogenic), whereas genistein is metabolised to the non-oestrogenic p-ethyl phenol. Inter-individual variation in ability to metabolise daidzein to equol could thus influence the potential health protective effects of soya isoflavones (see below). Unconjugated isoflavones are absorbed quickly from the upper small intestine in rats (Sfakisnos et al. 1997), whereas the glycoside conjugates are absorbed more slowly, which is consistent with their hydrolysis at more distal sites in the intestine to the unconjugated isoflavones (Barnes et al. 1996; King et al. 1996; King and Bursill, 1998). Once absorbed into the body, isoflavonoids are rapidly converted to their β-glucuronides (Sfakianos et al. 1997) and sulphate ester conjugates (Yasuda et al. 1996), it has not yet been established whether conjugated isoflavonoids have biological activity. However, this is an important possibility because steroid sulphate esters have a greater biological activity than their unconjugated equivalents.

The lignan phytoestrogen precursors in plant material, matairesinol and secisolariciresinol are present in foods as glycosides (Richard et al. 1996) and are converted by gut bacteria to the two main mammalian lignans (health factors) enterolactone and enterodiol respectively (Thompson 1994). Matairesinol undergoes dehydroxylation and demethylation directly to enterolactone, whereas secisolariciresinol is converted to enterodiol, which can then be oxidised to enterolactone (Adlercreutz 1996). After absorption, enterolactone and enterodiol are converted to their β-glucuronides (Fotsis *et al.* 1982) and eventually excreted in urine (see below), although relatively limited studies of the metabolism of lignans in humans have been carried out. A study in ileostomists found, in the absence of sufficient bacterial activity for conversion of plant lignans to mammalian lignans, no increase in plasma levels of mammalian lignans (Pettersson *et al.* 1996).

The urinary excretion of phytoestrogens is used as a measure of intake and thus the possible exposure of tissues to bioavailable phytoestrogens (which could confer protection against cancer). In assessing exposure to the protective effects of phytoestrogens, urinary excretion rates should be considered in combination with the plasma levels attained. A very low urinary excretion of lignans and equol has been observed in post-menopausal breast cancer patients compared to vegetarians (Adlercreutz *et al.* 1995; Adlercreutz 1996). In humans, omnivorous subjects usually have low levels of isoflavonoid excretion, and in individuals consuming a Western diet urinary levels of lignans have been found to be greater than those of isoflavones (Adlercreutz *et al.* 1995; Adlercreutz 1996). Urinary excretion of the lignans enterodiol and enterolactone was elevated in subjects consuming a carotenoid vegetable diet and a cruciferous vegetable diet compared to the basal (vegetable-free) diet, suggesting these vegetables may provide a source of mammalian lignan precursors (Kirkman *et al.* 1995). Moreover, men excreted less enterodiol and more enterolactone compared to women, suggesting there may be a gender difference in the bacterial metabolism of lignans in the colon (Kirkman *et al.* 1995). Urinary excretion of genistein, daidzein, equol and *O*-desmethylangolensin was found to be greater in subjects when they consumed a soya-rich diet compared to vegetable-rich (carotenoid versus cruciferous) and basal (vegetable-free) diets (Kirkman *et al.* 1995). Young American women from different ethnic groups have been found to predominantly excrete different types of phytoestrogens in their urine and presumably this reflects cultural influences on the diets consumed (Horn-Ross *et al.* 1997). Urinary isoflavonoid excretion has been reported to be dose-dependent at low to moderate levels of soya consumption (Karr *et al.* 1997).

In a study of plasma isoflavonoid levels in Japanese men compared to Finnish men, the means of the total daidzein, genistein, *O*-desmethylango-lensin, and equol levels were approximately 17-fold, 44-fold, 33-fold and 55-fold higher for the Japanese subjects compared to the Finnish ones (Adlercreutz *et al.* 1993). In a study of post-menopausal Australian women consuming soya flour, mean plasma levels of daidzein and equol of 68 ng/ml and 31 ng/ml, respectively were observed (Morton *et al.* 1994). In some subjects, however, levels as high as 148 ng/ml for genistein, 43 ng/ml for equol and 312 ng/ml for daidzein were reached (Morton *et al.* 1994). Moreover, only 33% of subjects were able to metabolize daidzein to equol. In addition, combined levels of enterolactone and enterodiol of

500 ng/ml were reached following linseed supplementation (Morton *et al.* 1994). In a study of the short-term metabolism of isoflavonoid and lignan phytoestrogen, male subjects ingested a cake containing soya flour and cracked linseed (Morton *et al.* 1997). Increased concentrations of both genistein and daidzein were observed within 30 minutes of consumption, with maximum concentrations reached by 5.5–8.5 hours following consumption. Maximum plasma concentrations were 87.5–190 ng/ml for genistein and 83.3–129 ng/ml for daidzein: none of the subjects were able to metabolise daidzein to equol. Metabolism of the plant lignans to the mammalian lignans was slower, with increased plasma levels of enterolactone and enterodiol found after 8.5 hours after consumption (Morton *et al.* 1997). Despite differences in urinary excretion, the bioavailabilities of genistein and daidzein, as determined by the areas under their plasma concentration-time curves (approximately equal) is similar: the fate of genistein is unclear as only 20% of the ingested dose is excreted in the urine (King and Bursill 1998). Chronic soya ingestion (as soya milk) appears to modulate the metabolism and disposition of ingested isoflavones in young females and an increase in the production of equol was observed in some of the subjects (Lu *et al.* 1996). In contrast, in males, chronic soya exposure did not alter the pathways of isoflavone metabolism, but altered the time courses of excretion (Lu *et al.* 1995).

Variable metabolic response to isoflavones has been reported for subjects who consumed soya flour over two days (Kelly *et al.* 1995). Urinary levels of genistein, daidzein, equol and O-desmethylangolensin were elevated 8-fold, 4-fold, 45-fold and 66-fold respectively, over baseline following consumption. Considerable inter-individual variation in metabolic response was found with the peak levels of equol showing the greatest variation (Kelly *et al.* 1995). The reasons for the wide inter-individual variation in isoflavone metabolite excretion following the consumption of soya have not yet been fully elucidated; however, in a recent study female equol excretor have been reported to consume a higher percentage of energy as carbohydrate and also greater amounts of plant protein and non-starch polysaccharide (NSP) (Lampe *et al.* 1998). In our randomised cross-over dietary intervention study in healthy male and female subjects, a diet high in isoflavones (soya-derived textured vegetable protein product containing 56 mg/day) or a diet low in isoflavones (soya-derived textured vegetable protein product containing 2 mg/day) were each consumed for two weeks separated by a four-week washout period. We found considerable inter-individual variation in metabolic response and that good equol producers (35% of subjects) consumed significantly less fat and more carbohydrate (also greater amounts of NSP) compared to the poor equol producers (Bowey *et al.* 1998; Rowland *et al.* 2000).

A recent epidemiological case–control study in breast cancer patients has shown that a high urinary excretion of both equol and enterolactone were found to be associated with a significant decrease in breast cancer risk (Ingram *et al*. 1997). The decrease in risk associated with daidzein excretion, however, was not significant (Ingram *et al*. 1997). This indicates the possible importance of phytoestrogen metabolism in decreased breast cancer risk, the relationship need not necessarily be causative as the presence of the phytoestrogens may just be a marker of dietary differences (Barnes 1998).

Metabolism by the gut microflora is an important factor influencing the disposition of chemicals in the gut and can result in activation of substances to more biologically active products. The bioavailability of soya isoflavones has been shown to depend on the gut microflora in women (Xu *et al*. 1995). Differences in faecal excretion greatly influenced isoflavone bioavailability: urinary recovery of ingested isoflavone phytoestrogens was more than twice as high in high versus low faecal excretors (Xu *et al*. 1995). Gut bacterial enzymes such as β-glucuronidases can hydrolyse isoflavone conjugates to aglycones (rapidly reabsorbed). However, intestinal microflora can also extensively metabolise and degrade isoflavones and other flavonoids, thus preventing their reabsorption from the colon. If an individual possesses bacteria that are not effective in isoflavone metabolism and degradation then more isoflavones would be absorbed, which would explain the positive association between high faecal isoflavones and greater total urinary recovery of isoflavones (Xu *et al*. 1995). The presence of different populations of microflora in the human gut may influence the bioavailability of soya isoflavone phytoestrogens. Our dietary intervention study with soya isoflavones has demonstrated considerable inter-individual variation in isoflavone phytoestrogen metabolism (Rowland *et al*. 2000). Studies on the types of human gut microflora involved in phytoestrogen bioavailability are needed. The metabolism of daidzein to equol by certain types of gut bacteria has important implications for daidzein bioavailability because equol producers are likely to have lower levels of bioavailable daidzein. Equol has its own particular health benefits and concerns (i.e. it is a more potent oestrogen and antioxidant).

Analysis of human breast milk has detected conjugates of genistein and daidzein following the consumption of roasted soya beans (Franke and Custer 1996; Franke *et al*. 1999). Thus concentrations of daidzein and genistein of 20–28 ng/ml and 8–14 ng/ml respectively, have been found in a subject who naturally consumed a diet rich in soya products and from another subject following a soya challenge. A preferential excretion was observed of the main metabolites (of daidzein), equol and *O*-desmethylangolensin over the parent compounds, compared to the patterns observed in urine and in faeces. This may be due to the higher isoflavone : metabolite ratio found in plasma compared to urine and faeces (milk is produced by secretory processes of blood).

Lower levels of prostate cancer are reported for vegetarians who consume large amounts of plant material (rich in lignans) and populations that consume large amounts of soya foods, see below. Levels of isoflavonoids and lignans have, therefore, been measured in prostatic fluid from males from different parts of the world (Morton *et al.* 1997) to determine the levels of bioavailable (and thus bioactive) phytoestrogens in body tissues. The highest mean levels of daidzein and its metabolite equol were found in males from Hong Kong (70 ng/ ml and 29.2 ng/ml) and China (24.2 ng/ml and 8.5 ng/ml), compared to those in males from the United Kingdom (11.3 ng/ml and 0.5 ng/ml) and Portugal (4.6 ng/ml and 1.27 ng/ml). In contrast, the highest levels of lignans were found in prostatic fluid from men from Portugal (enterolactone 162 ng/ml, and enterodiol 13.5 ng/ml). The levels of enterolactone and enterodiol were similar in prostatic fluid from men from the other countries investigated (enterolactone 20.3–32.9 ng/ml and enterodiol 1.6–6.6 ng/ml).

Health protection by phytoestrogens

Epidemiological evidence for protection against breast and prostate cancer

The incidence of breast and prostate cancer is much greater in Western countries than in Far Eastern ones, where there is an abundance of soya phytoestrogens in the diet (see reviews: Adlercreutz *et al.* 1995; Adlercreutz 1996; Cassidy 1996; Bingham *et al.* 1998; Barnes 1998). When people from Pacific Rim countries emigrate to the USA their risk of breast and prostate cancer increases. The increase in prostate cancer risk in men occurs in the same generation, whereas for women the increase in breast cancer risk is observed in the next generation (Shimizu *et al.* 1991). These changes in breast and prostate cancer risk have been mostly attributed to changes in diet, in particular the switch to a low soya Western diet (Messina *et al.* 1994). In countries such as Japan, Korea, China and Taiwan, the mean daily intake of soya products has been estimated to be in the range 10–50 g compared to only 1–3 g in the USA (Messina *et al.* 1994). Diets in South-east Asia are, however, becoming increasingly more like those in the West, with a lower soya content and there has been an increase in the incidence of breast cancer (Barnes 1998). Increased soya intake has been associated with a lowered risk of breast cancer in two out of four epidemiological studies that examined a wide range of dietary components in relation to breast cancer risk: no significant effect was observed in the other two studies (Messina *et al.* 1994; Barnes 1998). Two further studies that examined soya intake specifically, soya protein (Lee *et al.* 1991) intake and tofu intake (Wu *et al.* 1996).

Anticancer action of phytoestrogens

Studies using animal models provided the initial experimental evidence that soya can prevent breast cancer (Messina *et al.* 1994; Barnes 1995, 1997, 1998). Reports of results of 26 animal studies of experimental carcinogenesis in which diets containing soy or isolated and purified soybean isoflavonoids were used, have been reviewed (Messina *et al.* 1994). In 17 of these studies (65%) protective effects were reported the risk of cancer (incidence, latency or tumour number) was greatly reduced, and no studies reported that soya intake increased tumour development. In a rat model of breast cancer (7,12-dimethylbenz[a]anthracene induced), genistein administered in high doses by injection to young animals (neonatal and pre-pubertal) suppressed the number of mammary tumours observed over a six-month period by 50% and delayed the appearance of the tumours (Lamartiniere *et al.* 1995). This indicates the likely importance of the timing of exposure to the protective components of soya. Similar protection has also been obtained by giving genistein to the mothers through their diet (0.25 g/k feed), such that offspring were exposed to dietary genistein, from conception through to 21 days after birth (Fritz 1998). When genistein was administered to rats over the age of 35 days, only a 27% reduction of mammary tumours was observed, suggesting in this model where genistein is acting as a primary preventative agent, early life exposure is very important (Barnes 1998).

Biochanin A is found in certain subterranean clovers and can be converted to genistein by demethylation in the liver in addition to that in the breast. Biochanin A was a good anticancer agent when administered *after* the carcinogen (Gotoh *et al.* 1998a), suggesting the benefits of isoflavones other than genistein may not be restricted to early life exposure. Furthermore, the level of daily dietary biochanin A used in these experiments was 0.48–2.4 mg/kg, which approximates closely to the daily intake of isoflavones in the diets of south-east Asians (0.5–1.0 mg/ kg) (Barnes 1998). It is likely to be of considerable importance for women on standard tamoxifen therapy that the fermented soy food miso (contains mostly unconjugated isoflavones) and tamoxifen acted together to cause an additive reduction in the number of mammary tumours in the rat model (Gotoh *et al.* 1998b). In the mouse model of breast cancer (tamoxifen is oestrogenic in this model), genistein exposure *in utero* caused oestrogenic effects (more rapid mammary gland development and earlier vaginal opening) (Hilakivi-Clarke *et al.* 1998); however, further chemoprevention studies are needed to determine whether these oestrogenic changes would lead to an increased risk of breast cancer (Barnes 1998).

Mechanisms of anticancer action of phytoestrogens

In order to understand the anticancer action of phytoestrogens it is necessary to consider a range of possible mechanisms of action. Genistein is a potent and specific *in vitro* inhibitor of tyrosine kinase action in the autophosphorylation of the epidermal growth factor (EGF) receptor (Akiyama *et al.* 1987). The EGF receptor is overexpressed in many cancers, in particular those with the greatest ability for metastasis (Kim *et al.* 1996). Genistein is, therefore, used as a pharmacological tool to probe the actions of peptide hormones and drugs for tyrosine kinase dependency. It has been assumed that some of the anticancer effects of genistein are mediated via inhibition of tyrosine kinase activity; however, this is likely to be an oversimplification of the true *in vivo* situation (Barnes 1998), see below.

Phytoestrogens have biphasic effects on the proliferation of breast cancer cells in culture. At concentrations greater than $5\,\mu M$, genistein exhibits a concentration-dependent ability to inhibit both oestrogen-stimulated and growth factor-stimulated cell proliferation, and the genistein-mediated inhibition can be reversed by 17β-oestradiol (not found with other bioflavonoids) (So *et al.* 1997). Low concentrations of genistein can stimulate the growth of oestrogen receptor-positive MCF-7 cells, but only in the absence of any oestrogens (Wang and Kurzer 1997; Zava and Duwe 1997). Genistein does not, however, stimulate the growth of oestrogen receptor-negative breast cancer cells (Peterson and Barnes 1991; Wang *et al.* 1996), in these cell lines it only inhibits cell proliferation (similar IC_{50} values to in oestrogen receptor-positive cells (Peterson and Barnes 1991).

Equol is a potent stimulator of the expression of oestrogen-specific genes and is around 100-fold more effective than daidzein (Sathyamoorthy and Wang 1997). Genistein is a much better ligand for ERβ (see above) with a K_d of 0.4 nM than for the ERα (Kuiper *et al.* 1997). However, genistein behaves as an incomplete oestrogen in human kidney cells transiently expressing ERβ, suggesting that it may be a partial oestrogen antagonist in cells expressing ERβ (Barkhem *et al.* 1998): in cells transiently expressing ERα, genistein is a full agonist. Other forms of ERβ are being identified which may represent mRNA splicing variants (Moore *et al.* 1998; Maruyama *et al.* 1998) as these may lack the oestrogen-binding domain though retain DNA-binding ability to the oestrogen response element, they may cause antioestrogen effects (Barnes 1998). Indeed, in the ERβ–genistein complex, the AF-2 helix (H12) lies in a similar position to that induced by ER antagonists rather than adopting the distinctive "agonist" orientation (Pike *et al.* 1999).

Mechanisms other than those involving oestrogen receptors are likely to be involved in the inhibition of cell proliferation by genistein because genistein inhibits EGF-stimulated in addition to 17β-oestradiol-stimulated growth of MCF-7 cells (Peterson and Barnes 1996). Originally it was suggested that the inhibitory action of genistein on cell proliferation, involved effects on the autophosphorylation of the EGF receptor in membranes from A-431 cells (Akiyama *et al.* 1987). Indeed a number of studies have shown that exposure to genistein can reduce the tyrosine phosphorylation of cell proteins in whole cell lysates. Studies using cultured human breast and prostate cancer cells, however, have not confirmed that genistein has a direct effect on the autophosphorylation of the EGF receptor (Peterson and Barnes 1993, 1996). Furthermore, *in vivo* studies in male rats have shown that genistein decreases the amount of EGF receptor present in the (Dalu *et al.* 1998) indicating that the observed reduction in tyrosine phosphorylation may be a secondary effect of the influence of genistein on the expression or turnover of EGF receptor (Barnes 1998).

Many other mechanisms of action for genistein in particular have been suggested. These include, inhibition of DNA topoisomerases (Okura *et al.* 1998; Kondo *et al.* 1991), inhibition of cell cycle progression (Rauth *et al.* 1997; Kim *et al.* 1998), inhibition of angiogenesis (Fotsis *et al.* 1993, 1995; Kruse *et al.* 1997; see below), tumour invasiveness (Su *et al.* 1998; Yan and Han 1998), inhibition of enzymes involved in oestrogen biosynthesis (see above) (Adlercreutz *et al.* 1993; Makela *et al.* 1998), effects on the expression of DNA transcription factors c-fos and c-jun (Wei *et al.* 1996), on reactive oxygen species (see Chapter 1) (Wei *et al.* 1995) and on oxidative membrane damage (Wiseman *et al.* 1998; see below) and oxidative damage *in vivo* (Wiseman *et al.* 1999; see below) and on transforming growth factor-β (TGF-β) (Peterson *et al.* 1996; Kim *et al.* 1998; Sathyamoorthy *et al.* 1998; see below).

Angiogenesis, the formation of new blood vessels, is under normal conditions a vitally important process for reproductive function, development and wound repair, which is very tightly controlled. Many disease states, however, involve persistent and unregulated angiogenesis and the growth and metastasis of tumours is dependent on angiogenesis. Genistein is a potent inhibitor of angiogenesis *in vitro* (Fotsis *et al.* 1993). This ability of genistein to inhibit tumour cell proliferation and angiogenesis suggests that it could have therapeutic applications in the treatment of chronic neovascular diseases including solid tumour growth (Fotsis *et al.* 1995). Indeed, inhibition of neovascularisation of the eye by genistein has been reported (Kruse *et al.* 1997).

Antioxidant properties have been reported for isoflavones (Wei *et al.* 1995; Wiseman 1996; Wiseman *et al.* 1998). In model membrane systems, equol was a more effective antioxidant than genistein or the parent compound daidzein. Equol had an IC_{50} value for inhibition of membrane lipid peroxidation of 5 μM compared to 30 μM for tamoxifen and 8 μM for 4-hydroxytamoxifen (Wiseman *et al.* 1998). Plasma values as high as 0.2 μM have been reported for equol (Morton *et al.* 1994), and although this appears to be considerably less than that required *in vitro* for effective membrane protection against oxidative damage, membrane accumulation *in vivo* to achieve effective concentrations is likely. Furthermore, equol shows structural similarity to the tocopherols (Barnes 1998, see Chapter 1) and antioxidant action could contribute to anticancer ability because reactive oxygen species (see Chapter 1) could initiate signal transduction through the MAP-kinases (Wiseman and Halliwell 1996; Barnes 1998).

Genistein may also act through enhancement of transforming growth factor-β (TGFβ) (Peterson *et al.* 1996; Kim *et al.* 1998; Sathyamoorthy *et al.* 1998): this mechanism may be a link between the effects of genistein in a variety of chronic diseases (Barnes 1998) including atherosclerosis (see below) and hereditary haemorrhagic telangiectasia (the Osler–Weber–Rendu syndrome) in which defects in TGFβ have been characterised (Johnson *et al.* 1996).

Clinical studies

In a study aimed at evaluating the use of phytoestrogens as preventative agents for breast cancer, an isolated soya protein beverage providing 42 mg genistein and 27 mg daidzein per day was administered to healthy pre- and post-menopausal women (for six months with a three-month control period before and after) to determine its influence on breast cancer risk factors measurable in nipple aspirate fluid (NAF) (Petrakis *et al.* 1996). No change in NAF was observed in post-menopausal women. Premenopausal women, however, showed an increase in NAF volume, which persisted even after treatment with the soya beverage ended. This indicates the isoflavones were having an oestrogenic effect in the pre-menopausal women, although the situation here is quite complex since induction of apoptosis and/or increased differentiation of tumour cells in the breast both require an initial period of cell proliferation (Barnes 1998). This study provides some cause for concern that risk of pre-menopausal breast cancer may actually be enhanced by phytoestrogens, although clearly further studies are needed.

Clinical studies are under way in women with breast cancer to try and determine whether phytoestrogens are therapeutic or have secondary prevention effects (Barnes 1998), the results of these studies are awaited. Another important question relates to the likely interaction of phytoestrogens with tamoxifen treatment (see below): it appears likely from the strong additive protective effect of miso (good source of unconjugated isoflavones) with tamoxifen in the rat model of breast cancer (Gotoh *et al.* 1998b) that any interaction will be beneficial.

By contrast our randomised cross-over dietary intervention study in healthy male and female subjects aimed at examining the influence of soya isoflavones on oxidative damage *in vivo*, appeared to show beneficial effects. A diet high in isoflavones (soya-derived textured vegetable protein product containing 56 mg/day) or a diet low in isoflavones (soy-derived textured vegetable protein product containing 2 mg/day) were each consumed for two weeks separated by a four-week washout period. The biomarkers of oxidative damage measured included plasma isoprostane concentrations (8-epi-PGF$_{2\alpha}$ an F$_2$-isoprostane was measured: see Chapter 1), which showed a significant decrease, suggesting soya isoflavones may protect against the oxidative damage implicated in cancer, atherosclerosis and cardiovascular disease generally (Wiseman *et al.* 2000).

Protection by phytoestrogens against cardiovascular disease and osteoporosis

Phytoestrogens can also imitate the protective action of oestrogen on the cardiovascular system and on bone. This action may contribute to their reported protective ability against heart disease and osteoporosis and suggests their use as a natural supplement, as an alternative to hormone replacement therapy (HRT) in women (Clarkson *et al.* 1995, 1998).

Cardioprotective action of phytoestrogens

Lower incidence of heart disease has also been reported in populations consuming large amounts of soy products. Soy protein incorporated into a low fat diet can reduce cholesterol and LDL (low-density lipoprotein)– cholesterol concentrations and these effects are likely to be mediated by the oestrogenic isoflavonoids present (Anderson *et al.* 1995). Soya protein incorporated into a low fat diet can reduce cholesterol and LDL and raise high-density lipoprotein (HDL) and the oestrogenic isoflavones present are likely to contribute to this effect (Raines and Ross 1995).

The antioxidant action of phytoestrogens may contribute also to their cardioprotective properties. Isoflavones can protect LDL against oxidative modification *in vitro*. The IC$_{50}$ value for equol was 3 μM compared to 15 μM for tamoxifen and 1 μM for 4-hydroxytamoxifen (Wiseman *et al.* 1998), and although this is greater than the bioavailable level of equol of 0.2 μM that can be achieved (see above), nevertheless, bioaccumulation into LDL particles and into the lipophilic atheromal plaques associated with the arterial wall is likely, as previously suggested, for tamoxifen and 4-hydroxytamoxifen (Wiseman *et al.* 1993a). Indeed, in our own recent dietary intervention study, subjects consumed soya-derived textured vegetable protein ± isoflavones (see above) plasma isoprostanes showed a significant decrease and resistance of LDL to oxidation, which showed a significant increase (Wiseman *et al.* 2000). The increased resistance of LDL to oxidation, following consumption of soya products, is in agreement with other similar studies (Tikkanen *et al.* 1998). Although total plasma cholesterol and apolipoprotein B concentrations were unaffected by the isoflavone content of the diet, HDL and apolipoprotein AI concentrations were significantly higher following the high isoflavone dietary period (Dean *et al.* 1998). These observations suggest that soya isoflavones may protect against the oxidative damage implicated in atherosclerosis and cardiovascular disease, generally (Wiseman *et al.* 2000).

Genistein has a number of properties that suggest its potential as an antiatherogenic agent. *In vitro* genistein is a potent inhibitor of tyrosine kinase activity (see above) and thus it is able to block the action of growth factors such as PDGF (platelet derived growth factor), basic FGF (fibroblast growth factor) and other growth factors that work through tyrosine kinase action and are implicated in the growth of the atherosclerotic lesion (PDGF is postulated to play an important role in the intimal smooth muscle cell proliferation that forms part of the atherosclerotic process, Wilcox and Blumenthal 1995). In addition, genistein may reduce overall thrombosis associated with atherosclerosis by interfering with platelets and thrombin action. Blockade of tyrosine kinase activity by genistein prevents thrombin-induced platelet activation and aggregation *in vitro*. Thrombus formation at sites of vascular injury involves deposition of fibrin and platelets and depends on the activation of platelets followed by their deposition and aggregation. Thrombin formation at the site of injury results in the production of fibrin and further platelet activation and aggregation, and thus is important in the formation of blood clots. PDGF is released by activated platelets during thrombus formation, thus genistein has the potential additional benefit of inhibiting the action of this important growth factor.

Genistein can also inhibit the cell proliferation involved in lesion formation (Raines and Ross 1995). Genistein inhibits the proliferation of many vascular cells including vascular endothelial cells (see above) and also inhibits the atherosclerotically important process of angiogenesis. Neovascularisation is associated with advanced lesions, and genistein inhibits this function in cultured endothelial cells (Fotsis *et al.* 1993, 1995, see above) and may have similar effects *in vivo* on the vascular wall endothelial cells. Proliferation of smooth muscle cells is also important for the progression of atherosclerotic lesions and genistein has been shown to inhibit this process. The expression of leukocyte adhesion molecules ($\beta2$ integrins) after activation is likely to be of important in the initiation of lesion formation, and genistein has been shown to inhibit cell adhesion (Raines and Ross 1995).

Protection against osteoporosis by phytoestrogens

Osteoporosis is a chronic disease in which the bones become brittle and break more easily. Post-menopausal women, suffer hip fractures caused by osteoporosis, which develops primarily as a consequence of the low oestrogen levels that occur after the menopause (Thorneycroft 1989). Pre-menopausal women are therefore, protected by their oestrogen levels against osteoporosis. Although calcium supplementation is important before the menopause, on its own it cannot stop bone loss in peri-menopausal and post-menopausal women. HRT is very effective: a dose of 0.625 mg/day of conjugated oestrogens can prevent bone loss (Thorneycroft 1989). Furthermore, further studies have shown HRT to be osteoprotective if taken after the menopause for more than 24 months (Fentiman et al. 1994). Phytoestrogens may protect against post-menopausal bone loss and osteoporosis and studies to investigate this are in progress. Indeed, the drug ipriflavone (an isoflavone derivative) at a dose of 600 mg/day can prevent the increase in bone turnover and the decrease bone density in post-menopausal women (Gambacciani et al. 1997).

Phytoestrogens: the dangers versus future dietary benefits and use in nutraceutical supplements

Phytoestrogens cause infertility in some animals and concern have been raised over their consumption by human infants. The isoflavones found in a subterranean clover species (in Western Australia) have been identified as the agents responsible for an infertility syndrome in sheep (Adams 1995). In addition, isoflavones in the soya in the diets of cheetahs in captivity has been shown to lead to their infertility (Setchell 1987). Furthermore, phytoestrogens, including isoflavones, when fed at relatively high doses have been shown in some rodent studies to cause changes to the immature reproductive tract resulting in developmental abnormalities. Most animals that are bred commercially and domestic animals, however, are fed diets containing soya (up to 20% by weight) without any apparent reproductive problems (Barnes 1998). A study in peripubertal monkeys showed no adverse effects, and no reproductive abnormalities have been found in people living in countries where soya consumption is high. Indeed, the finding that dietary isoflavones are excreted into breast milk by soya-consuming mothers suggests that in cultures where consumption of soya products is the norm, breast-fed infants are exposed to high levels, again without any adverse effects. Isoflavone exposure shortly after birth at a critical developmental period through breastfeeding may protect against cancer and may be more important to the observation of lower cancer rates in populations in the Far East than adult dietary exposure to isoflavones (Franke et al. 1999). Controversy thus exists as to whether soya-based infant formulae containing isoflavones pose a health risk (Irvine et al. 1998). A recent review of the literature on the use of soya milk in infants, however, suggests that there is no real basis for concern (Klein 1998).

It is worth noting that epidemiological evidence suggesting protection from hormone-dependent cancer by phytoestrogens is based on foods rather than isoflavone extracts, and it is possible that soya foods should be included as a part of a healthy diet rather than reliance on isoflavone extracts. Furthermore, preparation of isoflavone extracts may result in the loss of important soya components that act synergistically with the isoflavones and may result in the daily dose of phytoestrogens being increased too far above the level that can be obtained from the diet and thus toxicity may be encountered: the levels for this are currently unknown. In addition, thorough investigations of isoflavone extracts (a number of these are now on the market) are clearly needed. Toxicity from isoflavones may arise from their action as alternative substrates for the enzyme thyroid peroxidase (Divi et al. 1997) thus potentially leading to the depletion of iodine and subsequent hypothyroidism, which has been suggested as a risk factor for breast cancer (Shering et al. 1996). People in South-east Asia would be protected by the inclusion of iodine-rich seaweed products in their diet. However, other bioflavonoids are a better substrate for this enzyme than isoflavones. If true this suggestion would mean that diets with the recommended five to seven daily servings of fruit and vegetables (rich sources of bioflavonoids and other polyphenols) would actually increase breast cancer risk rather than protect against it (Barnes 1998).

Whether we should include more foods containing phytoestrogens in our diet (and or as nutraceutical food supplements, see above) appears to be dependent on a number of factors including our age and sex. Certainly in terms of age, our biological receptiveness to the different potential protective (or even harmful) effects of phytoestrogens will change as we grow older. Menopausal women appear to be among those likely to benefit, gaining possible protection against heart disease and osteoporosis, caused by their oestrogen-deficient state. Older men may also benefit from protection against prostate problems and cancer, and heart disease. Indeed, it is significant that the Food and Drug Administration (USA) is now allowing health claims for protection against heart disease to be made for foods containing a certain amount of soya protein.

References

Adam, H.K., Patterson, J.S. and Kemp, J.V. (1980) Studies on the metabolism and pharmacokinetics of tamoxifen in normal volunteers. *Cancer Treat. Rep.*, **64**, 61–764.

Adams, N.R. (1995) Detection of the effects of phytoestrogens on sheep and cattle. *J. Anim. Sci.*, **73**, 1509–1515.

Adlercreutz, H. (1996) Lignans and isoflavonoids: epidemiology and possible role in prevention of cancer. In: *Natural Antioxidants and Food quality in Atherosclerosis and Cancer Prevention*, J.T. Kumpulainen and J.T. Salonen, eds. Royal Society of Chemistry, London, pp. 349–355.

Adlercreutz, H., Bannwart, C., Wahala, K. *et al.* (1993) Inhibition of human aromatase by mammalian lignans and isoflavonoid phytoestrogens. *J. of Steroid Biochem. and Mol. Biol.*, **44**, 147–153.

Adlercreutz, C.H.T., Goldin, B.R., Gorbach, S.L., Hockersted, K.A.V., Watanabe, S., Hamalainen, E.K., Markkanen, M.H., Makela, T.H., Wahala, K.T., Hase, T.A. and Fotsis, T. (1995) Soybean phytoestrogen intake and cancer risk. *J. Nutr.*, **125**, 757S–770S.

Adlercreutz, H., Markkanen, H. and Watanabe, S. (1993) Plasma concentrations of phytoestrogens in Japanese men. *Lancet*, **342**, 1209–1210.

Akiyama, T., Ishida, J., Nakagawa, S. *et al.* (1987) Genistein, a specific inhibitor of tyrosine-specific protein kinases. *J. of Biol. Chem.*, **262**, 5592–5595.

Anderson, J.W., Johnstone, B.M. and Cook-Newell, M.E. (1995) Meta-analysis of the effects of soy protein intake on serum lipids. *N. Engl. J. Med.*, **333**, 276–282.

Andersson, A., Hultberg, B., Brattstrom, L. and Isaksson, A. (1992) Decreased serum homocysteine in pregnancy. *Eur. J. Clin. Chem. Clin. Biochem.*, **30**, 377–379.

Anker, G., Lonning, P.E., Ueland, P.M., Refsum, H. and Lien, E.A. (1995) Plasma levels of the atherogenic amino acid homocysteine in postmenopausal women with breast cancer treated with tamoxifen. *Int. J. Cancer*, **60**, 365–368.

Bar, P.R. and Amelink, G.J. (1997) Protection against muscle damage exerted by oestrogen: hormonal or antioxidant action? *Biochem. Soc. Trans.*, **25**, 50–54.

Barkhem, T., Carlsson, B. and Nilsson, Y. (1998) Differential response of estrogen receptor α and estrogen receptor β to partial estrogen agonists/antagonists. *Mol. Pharmacol.*, **54**, 105–112.

Barnes, S. (1995) Effects of genistein in in vivo and in vitro models of cancer growth. *J. Nutr.*, **125**, 777S–783S.

Barnes, S. (1997) The chemopreventive properties of soy isoflavonoids in animal models of breast cancer. *Breast Cancer Res. and Treat.*, **46**, 169–179.

Barnes, S. (1998) Phytoestrogens and breast cancer. In: *Baillière's Clinical Endocrinology and Metabolism*, Vol. 12 (No. 4), H. Adlercreutz, ed. Baillière Tindall, London, pp. 559–579.

Barnes, S., Sfakianos, J., Coward, L. *et al.* (1996) Soy isoflavonoids and cancer prevention: underlying biochemical and pharmacological issues. In: *Dietary Phytochemicals and Cancer Prevention*, R. Biltrum, ed. Plenum Press, pp. 87–100.

Bingham, S.A., Atkinson, C., Liggins, J., Bluck, L. and Coward, A. (1998) Plant oestrogens: where are we now. *Br. J. Nutr.*, **79**, 393–406.

Bowey, E.A., Rowland, I.R., Adlercreutz, H., Sanders, T.A.B. and Wiseman, H. (1998) Inter-individual variation in soya metabolism: the role of habitual diet. *Proc. of the Nutr. Soc.*, **57**, 161A.

Bradlow, H.L., Davis, D.L., Lin, G., Sepkovic, D. and Tiwari, R. (1995) Effects of pesticides on the ratio of 16α/2-hydroxyesterone: a biologic marker of breast cancer risk. *Environ. Health Perspectives*, **103**, 147–150.

Brzozowski, A.M., Pike, A.C.W., Dauter, Z., Hubbard, R.E., Bonn, T., Engstrom, O., Ohman, L., Greene, G.L., Gustafsson, J-A. and Carlquist, M. (1977) Molecular basis of agonism and antagonism in the oestrogen receptor. *Nature*, **389**, 753–758.

Buckley, M.M.T. and Goa, K.L. (1989) Tamoxifen: A reappraisal of its pharmaco-dynamic and pharmacokinetic properties and therapeutic use. *Drugs*, **37**, 451–490.

Cassidy, A. (1996) Physiological effects of phyto-oestrogens in relation to cancer and other human health risks. *Proc. Nutr. Soc.*, **55**, 399–417.

Clarkson, T.B., Anthony, M.S. and Hughes, C.L. Jr (1995) Estrogenic soybean isoflavones and chronic disease: risks and benefits. *Trends in Endocrinology and Metabolism*, **6**, 11–16.

Clarkson, T.B., Anthony, M.S., Williams, J.K., Honore, E.K. and Cline, J.M. (1998) The potential of soyabean phytoestrogens for postmenopausal hormone replacement therapy. *Proc. Soc. for Exp. Bio. and Med.*, **217**, 365–368.

Coward, L., Barnes, N.C., Setchell, K.D.R. and Barnes, S. (1993) The antitumour isoflavones, genistein and daidzein, in soybean foods of American and Asian diets. *J. Agric. and Food Chem.*, **41**, 1961–1967.

Dalu, A., Haskell, J.F., Coward, L. and Lamartiniere, C.A. (1998) Genistein, a component of soy, inhibits the expression of the EGF and ErbB/Neu receptors in the rat dorsolateral prostate. *Prostate*, **37**, 36–43.

Daniel, P.C., Gaskell, J. Bishop, H., Campbell, C. and Nicholson, R. (1981) Determination of tamoxifen and biologically active metabolites in human breast tumour and plasma. *Eur. J. Cancer and Clin. Oncol.*, **17**, 1183–1189.

Davies, K. (1995) Breast cancer genes: further enigmatic variations. *Nature*, **378**, 762–763.

Dean, T.S., O'Reilly, J., Bowey, E., Wiseman, H., Rowland, I. and Sanders, T.A.B. (1998) The effects of soyabean isoflavones on plasma HDL concentrations in healthy male and female subjects. *Proc. of the Nutr. Soc.*, **57**, 123A.

De Vos, D., Guelen, P.J.M. and Stevenson, D. (1989) The bioavailability of Tamoplex (tamoxifen). Part 4. A parallel study comparing Tamoplex and four batches of Nolvadex in healthy male volunteers. *Meth. and Find. Exp. Clin. Pharmacol.*, **11**, 647–655.

Divi, R.L., Chang, H.C. and Doerge, D.R. (1997) Anti-thyroid isoflavones from soybean: isolation, characterization, and mechanisms of action. *Biochem. Pharmacol.*, **54**, 1987–1096.

Fentiman, I.S., Wang, D.Y., Allen, D.S., De Stavola, B.L., Moore, J.W., Reed, M.J. and Fogelman, I. (1994) Bone density of normal women in relation to endogenous and exogenous oestrogens. *Br. J. of Rheumatol.*, **33**, 808–815.

Fotsis, T., Heikkinen, R., Adlercreutz, H. *et al.* (1982) Capillary gas chromatographic method for the analysis of lignans in human urine. *Clin. Chim. Acta*, **121**, 361–371.

Fotsis, T., Pepper, M., Adlercreutz, H. *et al.* (1993) Genistein, a dietary-derived inhibitor of in vitro angio-genesis. *Proc. of the Nat. Acad. Sci. US*, **90**, 2690–2694.

Fotsis, T., Pepper, M., Aldercreutz, H. *et al.* (1995) Genistein, a dietary ingested isoflavonoid, inhibits cell proliferation and in vitro angiogenesis. *J. Nutr.*, **125** (3 supplement), 790S–797S.

Franke, A.A. and Custer, L.J. (1996) Daidzein and genistein concentrations in human milk after soy consumption. *Clin. Chem.*, **42**, 955–964.

Franke, A.A., Yu, M.C., Maskarinec, G., Fanti, P., Zheng, W. and Custer, L.J. (1999) Phytoestrogens in human biomatrices including breast milk. *Biochem. Soc. Trans.,* **27**, 308–318.

Fritz, W., Wang, J., Coward, L. and Lamartiniere, C.A. (1998) Dietary genistein: perinatal mammary cancer prevention, bioavailability and toxicity testing in the rat. *Carcinogenesis,* **19**, 2151–2158.

Fromson, J.M., Pearson, S. and Bramah, S. (1973) The metabolism of tamoxifen (ICI 46,474). Part II: In female patients. *Xenobiotica,* **3**, 711–714.

Fryer, R.H., Wilson, B.D., Gubler, D.M., Fitzgerald, L.A. and Rogers, G.M. (1993) Homocysteine, a risk factor for premature vascular disease and thrombosis, induces tissue factor activity in endothelial cells. *Arterioscler. Thromb.,* **13**, 1327–1333.

Furr, B.J.A. and Jordan, V.C. (1984) The pharmacology and clinical uses of tamoxifen. *Pharmac. Ther.,* **25**, 127–205.

Gamabacciani, M., Ciaponi, M., Cappagli, B., Piagessi, L. and Genazzani, A.R. (1997) Effects of combined low dose of the isoflavone derivative ipriflavone and estrogen replacement on bone mineral density and metabolism in postmenopausal women. *Maturitas,* **28**, 75–81.

Golbitz, P. (1995) Traditional soyfoods: processing and products. *J. Nutr.,* **125**, 570S–572S.

Gotoh, T., Yamada, K., Yin, H. *et al.* (1998a) Chemoprevention of N-nitroso-N-methylurea-induced rat mammary carcinogenesis by soy foods or biochanin A. *Jap. J. of Cancer REs.,* **89**, 137–142.

Gotoh, I., Yamada, K., Ito, A. *et al.* (1998b) Chemoprevention of N-nitroso-N-methylurea-induced rat mammary cancer by miso and tamoxifen, alone and in combination. *Jap. J. Cancer Res.,* **89**, 487–495.

Grainger, D.J. and Metcalfe, J.C. (1996) Tamoxifen: teaching an old drug new tricks? *Nature Med.,* **2**, 381–385.

Grainger, D.J., Witchell, C.M. and Metcalf, J.C. (1995) Tamoxifen elevates transforming growth factor-β and suppresses diet-induced formation of lipid lesions in mouse aorta. *Nature Med,* **1**, 1067–1073.

Grazzini, E., Guillon, G., Mouillac, B. and Zingg, H. (1998) Inhibition of oxytocin receptor function by direct binding of progesterone. *Nature,* **392**, 509–512.

Guelen, P.J.M., Stevenson, D., Brigss, R.J. and De Vos, D. (1987) The bioavailability of Tamoplex (tamoxifen). Part 2. A single dose cross-over study in healthy male volunteers. *Meth. and Find. Exptl. Clin. Pharmacol.,* **9**, 685–690.

Guetta, V., Lush, R.M., Figg, W.D., Waclawiw, M.A. and Cannon, R.O. (1995) Effects of the antiestrogen tamoxifen on low-density lipoprotein concentrations and oxidation in postmenopausal women. *Am. J. Cardiol.,* **76**, 1072–1073.

Hayden, M.R. and Reidy, M. (1995) Many roads lead to atheroma. *Nature Med.,* **1**, 22–23.

Hilakivi-Clarke, L., Cho, E. and Clarke, R. (1998) Maternal genistein exposure mimics the effects of oestrogen on mammary gland development in female mouse offspring. *Oncol. Rep.,* **5**, 609–615.

Hirano, K., Ogihara, T., Miki, M., Yauda, H., Tamai, H., Kawamura, N. and Mino, M. (1994) Homocysteine induces iron-catalysed lipid peroxidation of low-density lipoprotein that is prevented by alpha-tocopherol. *Free Radical Res.,* **21**, 267–276.

Horn-Ross, P.L., Barnes, S., Kirk, M., Coward, L., Parsonnet, J. and Hiatt, R.A. (1997) Urinary phytoestrogen levels in young women from a multiethnic population. *Cancer Epidemiol. Biomarkers and Prev.,* **6**, 339–345.

Iiri, T., Farfel, Z. and Bourne, H.R. (1998) G-protein diseases furnish a model for the turn-on switch. *Nature*, **394**, 35–38.

Irvine, C.H.G., Fitzpatrick, M.G. and Alexander, S.L. (1998) Phyto-estrogens in soy-based infant foods – concentrations, daily intake, and possible biological effects. *Proc. Soc. Exp. and Biol. Med.*, **217**, 247–253.

Jacolot, F., Simon, I., Dreano, Y., Beaune, P., Riche, C. and Berthou, F. (1991) Identification of the cytochrome P-450 IIIA family as the enzymes involved in the N-demethylation of tamoxifen in human liver microsomes. *Biochem. Pharmacol.*, **41**, 1911–1919.

Johnson, D.W., Berg, J.N., Baldwin, M.A. *et al.* (1996) Mutations in the activin receptor-like kinase 1 gene in hereditary haemorrhagic telangiectasia type 2. *Nature Gen.*, **13**, 189–195.

Jordan, V.C. (1997) Tamoxifen: the herald of a new era of preventative therapeutics. *J. Nat. Cancer Inst.*, **89**, 747–749.

Katzenellenbogen, J.A., O'Malley, B.W. and Katzenellenbogen, B.S. (1996) Tripartite steroid hormone receptor pharmacology: interaction with multiple effector sites as a basis for the cell- and promoter-specific action for these hormones. *Molec. Endocrinol.*, **10**, 119–131.

Karr, S.C., Lampe, J.W., Hutchins, A.M. and Slavin, J.L. (1997) Urinary isoflavonoid excretion in humans is dose dependent at low to moderate levels of soy-protein consumption. *Am. J. Clin. Nutr.*, **66**, 46–51.

Kellen, J.A. (ed.) (1996) *Tamoxifen: Beyond the Antioestrogen*. Birkhauser, Boston.

Kelly, G..E., Joannou, G.E., Reeder, A.Y., Nelson, C. and Waring, M.A. (1995) The variable metabolic response to dietary isoflavones in humans. *Proc. Soc. Exp. Biol. and Med.*, **208**, 40–43.

Kim, H., Peterson, T.G. and Barnes, S. (1998) Mechanisms of action of the soy isoflavone genistein: emerging role of its effects through transforming growth factor beta signaling pathways. *American J. of Clin. Nutr.*, **68**, 1418S–1425S.

Kim, J.W., Kim, Y.T., Kim, D.K. *et al.* (1996) Expression of epidermal growth factor receptor in carcinoma of the cervix. *Gynecol. Oncol.*, **60**, 283–287.

King, R.A. and Bursill, D.B. (1998) Plasma and urinary kinetics of the isoflavones daidzein and genistein after a single soy meal in humans. *Am. J. Clin. Nutr.*, **67**, 867–872.

King, R.A., Broadbent, J.L. and Head, R.J. (1996) Absorption and excretion of the soy isoflavone genistein in rats. *J. Nutr.*, **126**, 176–182.

Kirkman, L.M., Lampe, J.W., Campbell, D.R., Martini, M.C. and Slavin, J.L. (1995) Urinary lignan and isoflavonoid excretion in men and women consuming vegetable and soy diets. *Nutr. and Cancer*, **24**, 1–12.

Klein, K.O. (1998) Isoflavones, soy-based infant formulas, and relevance to endocrine function. *Nutr. Rev.*, **561**, 193–204.

Kondo, K., Tsuneizumi, K., Watanabe, T. and Oishi, M. (1991) Induction of in vitro differentiation of mouse embryonal carcinoma (F9) cells by inhibitors of topoisomerases. *Cancer Res.*, **51**, 5398–5404.

Kruse, F.E., Joussen, A.M., Fotsis, T. *et al.* (1997) Inhibition of neovascularization of the eye by dietary factors exemplified by isoflavonoids. *Ophthalmologie*, **94**, 152–156.

Kuiper, G.G.J.M., Carlsson, B., Grandien, K., Enmark, E., Haggblad, J., Nilsson, S. and Gustaffsson, J-A. (1997) Comparison of the ligand binding specificity and transcript tissue distribution of oestrogen receptors α and β. *Endocrinology*, **138**, 863–870.

Kuiper, G.G.J.M., Enmark, E., Pelto-Huikko, M., Nilsson, S. and Gustafsson, J-A. (1996) Cloning of a novel estrogen receptor expressed in rat prostate and ovary. *Proc. Natl. Acad. Sci. US*, **93**, 5925–5930.

Lamartiniere, C.A., Moore, J.B., Brown, N.A. et al. (1995) Genistein suppresses mammary cancer in rats. Carcinogenesis, 16, 2833–2840.

Lampe, J.W., Karr, S.C., Hutchins, A.M. and Slavin, J.L. (1998) Urinary equol excretion with a soy challenge: influence of habitual diet. Proc. Soc. Exp. Biol. and Med., 217, 335–339.

Lapcik, O., Hill, M., Hampl, R., Wahala, K. and Adlercreutz, H. (1998) Identification of isoflavonoids in beer. Steroids, 63, 14–20.

Lee, H.P., Gourley, L., Duffy, S.W. et al. (1991) Dietary effects on breast-cancer risk in Singapore. Lancet, 337 (8751), 11997–1200.

Li, S., Ting, N.S.Y., Zheng, L., Chen, P.-L., Ziv, Y., Shiloh, Y., Lee, E.Y.-H.P. and Lee, W.-H. (2000) Functional link of BRCA1 and ataxia telangiectasia gene product in DNA damage response. Nature, 406, 210–215.

Liehr, J.G. (1999) 4-Hydroxylation of oestrogens as a marker for mammary tumours. Biochem. Soc. Trans., 27, 318–323.

Lien, E.A., Solheim, E., Lea, O.A., Lundgren, S., Kvinnsland, S. and Ueland, P.M. (1989) Distribution of 4-hydroxy-N-desmethyltamoxifen and other tamoxifen metabolites in human biological fluids during tamoxifen treatment. Cancer Res., 49, 2175–2183.

Lien, E.A., Anker, G., Refsum, H., Ueland, P.M. and Lonning, P.E. (1997) Effects of hormones on the plasma levels of the atherogenic amino acid homocysteine. Biochem. Soc. Trans., 25, 33–35.

Lien, E.A., Solheim, E., Ueland, P.M. (1991) Distribution of tamoxifen and its metabolites in rat and human tissues during steady-state treatment. Cancer Res., 51, 4837–4844.

Love, R.R., Wiebe, D.A., Feyzi, J.M., Newcombe, P.A. and Chappell, R.J. (1994) Effects of tamoxifen on cardiovascular risk factors in postmenopausal women after 5 years of treatment. J. Natl. Cancer Inst., 86, 1534–1539.

Love, R.R., Wiebe, D.A., Newcombe, P.A., Cameron, L., Leventhal, H., Jordan, V.C., Feyzi, J. and DeMets, D.L. (1991) Effects of tamoxifen on cardiovascular risk factors in postmenopausal women. Ann. Intern. Med., 115, 860–864.

Love, R.R., Surawicz, T.S. and Williams, E.C. (1992) Antithrombin III level, fibrinogen level, and platelet count changes with adjuvant tamoxifen therapy. Arch. Int. Med., 152, 317–320.

Lu, L-J.W., Lin, S-N., Grady, J.J., Nagamani, M. and Anderson, K.E. (1996) Altered kinetics and extent of urinary daidzein and genistein excretion in women during chronic soya exposure. Nutr. and Cancer, 26, 289–302.

Lu, L-J.W., Grady, J.J., Marshall, M.V., Ramanujam, S. and Anderson, K.E. (1995) Altered time course of urinary daidzein and genistein excretion during chronic soya diet in healthy male subjects. Nutr. and Cancer, 24, 311–323.

Makela, S., Poutanen, M., Kostlan, M.L. et al. (1998) Inhibition of 17-beta-hydroxysteroid oxidoreductase by flavonoids in breast and prostate cancer cells. Proc. Soc. Exp. Biol. and Med., 217, 310–316.

Mani, C., Gelboin, H.V., Park, S.S., Pearce, R., Parkinson, A. and Kupfer. D. (1993) Metabolism of the antimammary cancer antioestrogenic agent tamoxifen 1. Cytochrome P-450-catalysed N-demethylation and 4-hydroxylation. Drug Metab. Dispos., 21, 645–656.

Mani, C., Pearce, R., Parkinson, A. and Kupfer, D. (1994) Involvement of cytochrome P4503A in catalysis of tamoxifen activation and covalent binding to rat and human liver microsomes. Carcinogenesis, 15, 2715–2720.

Maruyama, K., Endoh, H., Sasakiiwaoka, H. et al. (1998) A novel isoform of rat estrogen receptor beta with 18 amino acid insertion in the ligand inding domain as a putative dominant negative regulator of estrogen action. *Biochem. and Biophys. Res. Commun.*, **246**, 142–147.

McDonald, C.C. and Stewart, H.J. for the Scottish Breast cancer Committee (1991) Fatal myocardial infarction in the Scottish adjuvant tamoxifen trial. *Br. Med. J.*, **303**, 435–437.

McVie, J.G., Simonetti, G.P.C., Stevenson, D., Briggs, R.J., Guelen, P.J.M. and De Vos, D. (1986) The bioavailability of Tamoplex (tamoxifen). Part 1. A pilot study. *Meth. and Find. Exptl. Clin. Pharmacol.*, **8**, 505–512.

Messina, M., Persky, V., Setchell, K.D.R. and Barnes, S. (1994) Soy intake and cancer risk: a review of the in vitro and in vivo data. *Nutr. and Cancer*, **21**, 113–131.

Miksicek, R.J. (1995) Estrogenic flavonoids: structural requirements for biological activity. *Proc. Soc. for Exp. Biol. and Med.*, **208**, 44–50.

Mitlak, B.H. and Cohen, F.J. (1997) In search of optimal long-term female hormone replacement: the potential of selective estrogen receptor modulators. *Hormone Res.*, **48**, 155–163.

Moore, J.T., Mckee, D.D., Slentzkesler, K. *et al.* (1998) Cloning and characterization of human estrogen receptor beta isoforms. *Biochem. and Biophys. Res. Commun.*, **247**, 75–78.

Morton, M.S., Matos-Ferreira, A., Abranches-Monteiro, L., Correia, R., Blacklock, N., Chan, P.S.F., Cheng, C., Lloyd, S., Chieh-ping, W. and Griffiths, K. (1997) Measurement and metabolism of isoflavonoids and lignans in the human male. *Cancer Lett.*, **114**, 145–151.

Morton, M.S., Wilcox, G., Wahlqvist, M.L. and Griffiths, K. (1994) Determination of lignans and isoflavonoids in human female plasma following dietary supplementation. *J. Endocrinol.*, **142**, 251–259.

Murphy, C., Fotsis, T., Pantzer, P., Adlercreutz, H. and Martin, F. (1987) Analysis of tamoxifen, *N*-desmethyltamoxifen and 4-hydroxytamoxifen levels in cytosol and KCl-nuclear extracts of breast tumours from tamoxifen treated patients by gas chromatography—mass spectrometry (GC–MS) using selected ion monitoring (SIM). *J. Steroid Biochem.*, **28**, 609–618.

Okura, A., Arakawa, H., Oka, H. *et al.* (1988) Effect of genistein on topoisomerase activity and on the growth of [Val 12]Ha-ras-transformed NIH 3T3 cells. *Biochem. and Biophys. Res. Commun.*, **157**, 183–189.

Pasagian-Macaulay, A., Meilahn, E.N., Bradlow, H.L., Sepkovic, D.W., Buhari, A.M., Simkin-Silverman, L., Wing, R.R. and Kuller, L.H. (1996) Urinary markers of estrogen metabolism 2- and 16-α-hydroxylation in premenopausal women. *Steroids*, **61**, 461–467.

Peeters, P.H., Verbeek, A.L., Krol, A. *et al.* (1995) Age at menarche and breast cancer risk in nulliparous women. *Breast Cancer Res. and Treat.*, **33**, 55–61.

Peterson, T.G. and Barnes, S. (1991) Genistein inhibition of the growth of human breast cancer cells: independence from estrogen receptors and the multi-drug resistance gene. *Biochem. and Biophys. Res. Commun.*, **179**, 661–667.

Peterson, T.G. and Barnes, S. (1993) Isoflavones inhibit the growth of human prostate cancer cell lines without inhibiting epidermal growth factor receptor autophosphorylation. *Prostate*, **22**, 335–345.

Peterson, T.G. and Barnes, S. (1996) Genistein inhibits both estrogen and growth factor simulated proliferation of human breast cancer cells. *Cell Growth and Differentiation*, **71**, 1345–1351.

Peterson, T.G., Barnes, S. and Kim, H. (1996) Mechanisms of action of the soy isoflavone genistein at the cellular level. *2nd International Symposium of the Role of Soy in the Prevention and Treatment of Chronic Diseases*, Brussels, Belgium.

Petrakis, N., Barnes, S., King, E.B. *et al.* (1996) Stimulatory influence of soy protein isolate on breast secretion in pre- and postmenopausal women. *Cancer Epidemiol: Biomarkers and Prevention*, **5**, 785–794.

Pettersson, D., Aman, P., Knudesen, K.E.B., Lundin, E., Zhang, J.X., Halimans, G., Harkonen, H. and Adlercreutz, H. (1996) Intake of rye bread by ileostomists increases ileal excretion of fiber polysaccharide components and organic acids but does not increase plasma or urine lignans and isoflavonoids. *J. Nutri.*, **126**, 1594–1600.

Pike, A.C.W., Brzozowski, A.M., Hubbard, R.E., Bonn, T., Thorsell, A.-G., Engström, O., Ljunggren, J., Gustafsson, J.-A. and Carlquist, M. (1999) Structure of the ligand-binding domain of oestrogen receptor beta in the presence of a partial agonist and a full antagonist. *The EMBO Journal*, **18**, 4608–4618.

Purohit, A., Sing, A. and Reed, M.J. (1999) Regulation of steroid sulphatase and oestradiol 17β-hydroxysteroid dehydrogenase in breast cancer. *Biochem. Soc. Trans.*, **27**, 323–327.

Raines, E.W. and Ross, R. (1995) Biology of atherosclerotic plaque formation: possible role of growth factors in lesion development and the potential impact of soy. *J. Nutr.*, **125**, 624S–630S.

Raman, V., Martensen, S.A., Reisman, D., Evron, E., Odenwald, W.F., Jaffee, E., Marks, J. and Sukumar, S. (2000) Compromised HOXA5 function can limit p53 expression in human breast tumours. *Nature* **405**, 974–978.

Rauth, S., Kichina, J. and Green, A. (1997) Inhibition of growth and induction of differentiation of metastatic melanoma cells in vitro by genistein: chemosensitivity is regulated by cellular p53. *Br. J. Cancer*, **75**, 1559–1566.

Reini, K. and Block, G. (1996) Phytoestrogen content of foods – a compendium of literature values. *Nutr. and Cancer*, **26**, 123–148.

Richard, S.E., Orcheson, L.J., Seidl, M.M. *et al.* (1996) Dose-dependent production of mammalian lignans in rats and in vitro from the purified precursor secoisolariciresinol diglycoside in flaxseed. *J. Nutr.*, **126**, 2012–2019.

Rowland, I., Wiseman, H., Sanders, T., Adlercreutz, H. and Bowey, E. (2000) Interindividual variation in metabolism of soy isoflavones and lignans: influence of habitual diet on equal production by the gut microflora. *Nutrition and Cancer*, **36**, 27–32.

Ruehlmann, D.R. and Mann, G.E. (1997) Actions of oestrogen on vascular endothelial and smooth-muscle cells. *Biochem. Soc. Trans.*, **25**, 40–45.

Russo, J. and Russo, I.H. (1995) The etiopathogenesis of breast cancer prevention. *Cancer Lett.*, **90**, 81–89.

Rutqvist, L.E. and Mattsson, A. (1993) Cardiac and thromboembolic morbidity among postmenopausal women with early-stage breast cancer in a randomized trial of adjuvant tamoxifen. *J. Natl. Cancer Inst.*, **85**, 1398–1406.

Sack, M.N., Rader, D.J. and Cannon, R.O. (1994) Oestrogen and inhibition of oxidation of low-density lipoproteins in postmenopausal women. *Lancet*, **343**, 269–270.

Sathyamoorthy, N. and Wang, T.T. (1997) Differential effects of dietary phyto-oestrogens daidzein and equol on human breast cancer MCF-7 cells. *Europ. J. Cancer*, **33**, 2384–2389.

Sathyamoorthy, N., Gilsdorf, J.S. and Wang, T.T.Y. (1998) Differential effect of genistein on transforming growth factor beta-1 expression in normal and malignant mammary epithelial cells. *Anticancer Res.*, **18**, 2449–2453.

Schapira, D.V., Kumar, N.B. and Lyman, G.H. (1990) Serum cholesterol reduction with tamoxifen. *Breast Cancer Res. and Treat.*, **17**, 3–7.

Setchell, K.D., Gosselin, S.J., Welsh, M.B. *et al.* (1987) Dietary estrogens – a probable cause of infertility and liver disease in captive cheetahs. *Gastroenterology*, **93**, 225–233.

Sfakianos, J., Coward, L., Kirk, M. and Barnes, S. (1997) Intestinal uptake and biliary excretion of the isoflavone genistein in the rat. *J. Nutr.*, **127**, 1260–1268.

Shering, S.G., Zbar, A.P., Moriarty, M. *et al.* (1996) Thyroid disorders and breast cancer. *Eur. J. of Cancer Preven.*, **5**, 504–506.

Shewmon, D.A., Stock, J.L., Rosen, C.J., Heiniluoma, K.M., Hogue, M.M., Morrison, A., Doyle, E.M., Ukena, T., Weale, V. and Baker, S. (1994) Tamoxifen and estrogen lower circulating lipoprotein(a) concentrations in healthy postmenopausal women. *Arterioscler. Thromb.*, **14**, 1586–1593.

Shmizu, H., Ross, R.K., Bernstein, L. *et al.* (1991) Cancers of the prostate and breast among Japanese and white immigrants in Los Angeles County. *Br. J. Cancer*, **63**, 963–966.

Slavin, J., Jacobs, D. and Marquart, L. (1997) Whole-grain consumption and chronic disease: Protective mechanisms. *Nutr. and Cancer*, **27**, 14–21.

So, F.V., Guthrie, N., Chambers, A.F. and Carroll, K.K. (1997) Inhibition of proliferation of estrogen receptor-positive MCF-7 human breast cancer cells by flavonoids in the presence and absence of excess estrogen. *Cancer Lett.*, **112**, 127–133.

Steinberg, D. and Lewis, A. (1997) Oxidative modification of LDL and atherogenesis. *Circulation*, **95**, 1062–1071.

Steinberg, D., Parthasarathy, S., Carew, T.E., Khoo, J.C. and Witzum, J.L. (1989) Modifications of low-density lipoprotein that increase its atherogenicity. *New England J. Med.*, **320**, 915–924.

Su, S.M., Dibattista, J.A., Sun, Y. *et al.* (1998) Up-regulation of tissue inhibitor of metalloproteinases-3 gene expression by TGF-beta in articular chondrocytes is mediated by serine/threonine and tyrosine kinases. *J. Cell. Biochem.*, **70**, 517–527.

Tikkanen, M.J., Wahala, K., Ojala, S., Vihma, V. and Adlercreutz, H. (1998) Effect of soybean phytoestrogen intake on low density lipoprotein resistance. *Proc. Nat. Acad. Sci.*, **95**, 3106–3110.

Thomas, H.V., Key, T.J., Allen, D.S., Moore, J.W., Dowsett, M., Fentiman, I.S. and Wang, D.Y. (1997a) A prospective study of endogenous serum hormone concentrations and breast cancer risk in post-menopausal women on the island of Guernsey. *Br. J. Cancer*, **76**, 401–405.

Thomas, H.V., Reeves, G.K. and Key, T.J.A. (1997b) Endogenous estrogen and postmenopausal breast cancer: a quantitative review. *Cancer Causes and Control*, **8**, 922–928.

Thompson, L.U. (1994) Antioxidant and hormone-mediated health benefits of whole grains. *Crit. Rev. in Food Sci. and Nutr.*, **34**, 473–497.

Thorneycroft, I.H. (1989) The role of oestrogen replacement therapy in the prevention of osteoporosis. *Am. J. Obst. and Gynecol.*, **160**, 1306–1310.

Wang, C. and Kurzer, M.S. (1997) Phytoestrogen concentration determines effects on DNA synthesis in human breast cancer cells. *Nutr. and Cancer*, **28**, 236–247.

Wang, T.T.Y., Sathyamoorthy, N. and Phang, J.M. (1996) Molecular effects of genistein on estrogen receptor-mediated pathways. *Carcinogenesis*, **17**, 271–275.

Wei, H., Barnes, S. and Wang, Y. (1996) Inhibitory effect of genistein on a tumor promotor-induced c-fos and c-jun expression in mouse skin. *Oncol. Rep.*, **3**, 125–128.

Wei, H., Bowen, R., Cai, Q. *et al.* (1995) Antioxidant and antipromotional effects ot the soybean isoflavine genistein. *Proc. Soc. for Exp. Biol. and Med.*, **208**, 124–130.

Westerveld, H.E., de Bruin, T.W.A. and Erkelens, D.W. (1997) Oestrogens and postprandial lipid metabolism. *Biochem. Soc. Trans.*, **25**, 45–49.

Wilcox, J.N. and Blumenthal, B.F. (1995) Thrombotic mechanisms in atherosclerosis: potential impact of soy proteins. *J. Nutr.*, **125**, 631S–638S.

Williams, J.K., Wagner, J.D., Li, Z., Golden, D.L. and Adams, M.R. (1997) Tamoxifen inhibits arterial accumulation of LDL degradation products and progression of coronary artery atherosclerosis in monkeys. *Arterioscler. Thromb. and Vasc. Biol*, **17**, 403–408.

Wiseman, H. (1994) *Tamoxifen: Molecular Basis of use in Cancer Treatment and Prevention.* John Wiley, Chichester.

Wiseman, H. (1996a) Dietary influences on membrane function: importance in protection against oxidative damage and disease. *J. Nutr. Biochem.*, **7**, 2–15.

Wiseman, H. (1996b) Tamoxifen: absorption, distribution, metabolism and excretion. In: *IARC Monographs on the evaluation of the carcinogenic risks to humans*, Vol. 66, *Some Pharmaceutical Drugs*, pp. 291–296.

Wiseman, H. (1996c) Role of dietary phyto-oestrogens in the protection against cancer and heart disease. *Biochem. Soc. Trans.*, **24**, 795–800.

Wiseman, H. (1998) Phytochemicals (b) Epidemiological Factors. In: *Encyclopedia of Human Nutrition*, M. Sadler, B. Caballero and S. Strain, eds. Academic Press, London, pp. 1549–1561.

Wiseman, H. (1999a) Importance of oestrogen, xenoestrogen and phytoestrogen metabolism in breast cancer risk. *Biochem. Soc. Trans.*, **27**, 299–304.

Wiseman, H. (1999b) The bioavailability of non-nutrient plant factors: dietary flavonoids and phyto-oestrogens. *Proc. Nutr. Soc.*, **58**, 139–146.

Wiseman, H., Arnstein, H.R.V., Cannon, M. and Halliwell, B. (1990) Mechanism of inhibition of lipid peroxidation by tamoxifen and 4-hydroxytamoxifen introduced into liposomes: Similarity to cholesterol and ergosterol. *FEBS Lett.*, **274**, 107–110.

Wiseman, H. and Halliwell, B. (1996) Damage to DNA by reactive oxygen and nitrogen species: role in inflammatory disease and progression to cancer. *Biochem. J.*, **313**, 17–29.

Wiseman, H. and O'Reilly, J. (1997) Oestrogens as antioxidant cardioprotectants. *Biochem. Soc. Trans.*, **25**, 54–59.

Wiseman, H., O'Reilly, J.D., Adlercreutz, H., Mallet, A.I., Bowey, E.A., Rowland, I.R. and Sanders, T.A.B. (2000) Isoflavone phytoestrogens consumed in soy decrease F_2-isoprostane concentrations and increase resistance of low-density lipoprotein to oxidation in humans. *Am. J. Clin. Nutr.* (in press).

Wiseman, H., O'Reilly, J., Lim, P., Garnett, A.P., Huang, W-C. and Sanders, T.A.B. (1998) Antioxidant properties of the isoflavone phytoestrogen functional ingredient in soya products. In: *Functional Foods, the Consumer, the Products and the Evidence*, M. Sadler and M. Saltmarsh, eds. Royal Society of Chemistry, Cambridge, pp. 80–86.

Wiseman, H., Paganga, G. Rice-Evans, C. and Halliwell, B. (1993) Protective actions of tamoxifen and 4-hydroxytamoxifen against oxidative damage to human low-density lipoproteins: a mechanism accounting for the protective action of tamoxifen? *Biochem. J.*, **292**, 635–638.

Wiseman, H., Quinn, P. and Halliwell, B. (1993b) Tamoxifen and related compounds decrease membrane fluidity in liposomes. Mechanism for the antioxidant action of tamoxifen and relevance to its anticancer and cardioprotective actions? *FEBS Lett.*, **330**, 53–56.

Witztum, J.L. (1994) The oxidation hypothesis of atherosclerosis. *Lancet*, **344**, 793–795.

Wu, A.H., Ziegler, R.G. and Horn-Ross, P.L. (1996) Tofu and risk of breast cancer in Asian-Americans. *Cancer Epidemiol. Biomarkers and Prev.*, **5**, 901–906.

Xu, X., Harris, K.S., Wang, H-J., Murphy, P.A. and Hendrich, S. (1995) Bioavailability of soybean isoflavones depends upon gut microflora in women. *J. Nutr.*, **125**, 2307–2315.

Yan, C.H. and Han, R. (1998) Genistein suppresses adhesion-induced protein tyrosine phosphorylation and invasion of B16-B16 melanoma cells. *Cancer Lett.*, **129**, 117–1224.

Yasuda, T., Mizunuma, S., Kano, Y. *et al.* (1996) Urinary and biliary metabolites of genistein in rats. *Biol. and Pharm. Bull.*, **19**, 413–417.

Yin, Y., Terauchi, Y., Solomon, G.G., Aizawa, S., Rangarajan, P.N., Yazaki, Y., Kadowaki, T. and Barrett, J.C. (1998) Involvement of p85 in p053-dependent apoptotic response to oxidative stress. *Nature*, **391**, 707–710.

Zava, D.T. and Duwe, G. (1997) Estrogenic and antiproliferative properties of genistein and other isoflavonoids in human breast cancer cells in vitro. *Nutr. and Cancer*, **27**, 31–40.

7 Strategies for the Removal of Ecotoxicants: Environmental Oestrogens and Oestrogen-mimics

TIM RIDGWAY[1] AND HELEN WISEMAN[2]

[1]Wise Associates, School of Biological Sciences, University of Surrey, Guildford, Surrey GU2 7XH

[2]Nutrition, Food and Health Research Centre, King's College London, Franklin-Wilkins Building, 150 Stamford Street, London SE1 8WA

Oestrogen mimicry in the environment

There is controversy over whether a decline in male reproductive health is occurring. Oestrogens, oestrogen-mimics and other endocrine disrupting substances, have been demonstrated to cause significant adverse effects on aquatic wildlife (Environment Agency Report 1996, 1998; MRC/IEH report 1995). The offending oestrogenic pollutants/toxicants include natural sterols, (most significantly 17β-oestradiol), and synthetic oestrogenic-mimics (xenoestrogens), for example, phthalates and alkyphenols and it is this diverse range of compounds which show the oestrogenic activity, which consequently results in oestrogenic substances being found in the environment, in food packaging and in foodstuffs (Wiseman et al., 1998a,b; Ridgway and Wiseman 1998). The toxic equivalency factor approach could be used for the hazard and risk assessment of such chemical mixtures (Safe 1998). Due to the widespread distribution of xenoestrogens methodologies are needed to determine the exact levels of exposure (Jimenez 1997). Human exposure to xenoestrogens also need to be carefully assessed and appropriate testing systems are needed to screen for alterations in endocrine activities, together with appropriate biomarkers of cumulative exposure (Rivas et al. 1997). If male reproductive health is at risk and needs to be protected then it is likely that total human exposure to oestrogens will have to be biomonitored. A combination of approaches may then be necessary to lower total exposure. These include the prevention of use of some classes of oestrogenic compound, the removal by filtration or bioremediation of offending chemicals and modification to inactive forms. The technology which could enable reduced oestrogen exposure, ranges from activated-carbon filtration to the genetic engineering of plants (Ridgway and Wiseman 1998), see also Figures 7.1, 7.2, 7.3 and 7.4.

The biomolecular basis of oestrogenicity

Oestrogens are involved in the growth, development and homeostasis of particular tissues (Ciocco and Roig 1995). The biological activity of oestrogens is classically considered to be mediated via the oestrogen receptor (ER), which is a ligand-inducible nuclear transcription factor (Tsai and O'Malley 1994). The binding of oestrogen to the ligand-binding domain of the ER triggers a sequence of molecular interactions resulting in either the activation or repression of target genes. The direct interaction of the ER with other components of the transcriptional system mediates transcriptional regulation (Katzenellenbogen *et al.* 1996). The crystal structures of the ligand-binding domain of the ER complexed to either the endogenous oestrogen 17β-oestradiol or to the selective antagonist raloxifene (structurally related to tamoxifen) has been reported (Brzozowski *et al.* 1997). Raloxifene inhibits the mitogenic effects of oestrogen in reproductive tissues maintaining the beneficial effects of oestrogen in other tissues. These crystal structures provide a molecular basis for agonism and antagonism in the oestrogen receptor: both agonist and antagonist bind at the same site within the core of the ligand-binding domain but each can induce a unique conformation in the transactivation domain of the ER (Brzozowski *et al.* 1997).

The 17β-oestradiol binding cavity is isolated from the external environment and comprises a high proportion of the hydrophobic core of the ER ligand-binding domain, and recognition is achieved via a mixture of specific hydrogen bonds, the complementary shape of the binding cavity and the polar nature of 17β-oestradiol (Brzozowski *et al.* 1997). A combination of specific polar and non-polar interactions thus enables the ER to selectively recognize and bind 17β-oestradiol with subnanomolar affinity in preference to a wide range of other endogenous steroids. The oestrogen receptor is the only steroid receptor able to interact additionally with a large number of non-steroidal compounds, and although the molecular fit of the binding cavity around the A-ring of 17β-oestradiol indicates that effective ligands must possess an aromatic ring, a wide range of different hydrophobic groups can be accepted by the remainder of the binding pocket (Brzozowski *et al.* 1997). The switch from agonism to antagonism appears to involve a ligand-mediated repositioning of helix 12 of the ligand-binding domain of the ER such that a transcriptionally competent transcriptional activation function (AF-2) is no longer generated. Phytoestrogens and environmental xenoestrogens demonstrate a structural similarity to the steroid nucleus of oestrogen: in particular a phenolic ring analogous to ring A in oestradiol. This allows them to bind to oestrogen and androgen receptors to produce effects ranging from agonism to antagonism of the endogenous hormone ligand (Miksicek 1995).

An oestrogenic substance may be defined as one which binds to an oestrogen receptor, for example the human oestrogen receptor (hER), and induces transcription. Endogenous oestrogenic substances are sterols; the basis of their oestrogenic activity has been attributed to a p-monophenolic group with an additional hydroxyl group 12 Å away. The oestrogen receptor is the only steroid receptor able to additionally interact with a large number of non-steroidal compounds and recent reports have indicated that although the molecular fit of the binding cavity around the A-ring of 17β-oestradiol appears to require that all effective ligands possess an aromatic ring, a wide range of different hydrophobic groups can be accepted by the remainder of the binding pocket (Brzozowski et al. 1997). This explains why such a range of substances have been found to possess oestrogenic activity. What is also apparent is the scale of the problem of removing oestrogenic substances from the environment and the difficulty in developing non-oestrogenic alternatives to many industrial chemicals. To remove the oestrogenicity of environmental substances through structural modulation would require addition of further hydroxyl groups to the principal p-monophenolic ring, because this has been observed to result in the removal or drastic reduction of oestrogenic activity (Miksicek 1995). The addition of hydroxyl groups may lead, however, to compounds which, although not toxic in terms of oestrogenicity, are potentially toxic by other mechanisms.

Identification of environmental xenoestrogens as oestrogen-mimics and their effects

Oestrogen-like biological activity in male fish, alligators, polar bears and seabirds can be caused by environmental compounds or their degradation products acting as oestrogen-mimics in the water environment (Environment Agency Report 1996, 1998; MRC/IEH report, 1995; Wiseman et al. 1998a,b; Ridgway and Wiseman 1998). Some of these environmental oestrogen-mimics (Olea et al. 1998; Pazos et al. 1998) derive from the biodegradation of domestic or industrial chemicals discharged into the environment. Examples include nonylphenol derivatives (and octylphenols) resulting from bacterial action on domestic detergents and the plant sterol β-sitosterol present in relatively large amounts in most pulp and paper mill effluents, while bisphenol A (4,4-isopropylidenediphenol) is derived from industrial resins and phthalates are present in some wrapping plastics (especially in polyvinylchloride). In addition, styrenes can leach into fatty foods from polystyrene food packaging.

Exposure of young males of the common carp (Cyprinus carpio) to 4-tert-pentylphenol during the period of sexual differentiation results in feminisation (Gimeno et al., 1998a) and exposure during spermatogenesis results in demasculinisation (Gimeno *et al.* 1998b). 4-nonylphenol (and 17β-oestradiol) inhibit the settlement of the cypris larva of the barnacle, *Balanus amphitrite* (Billinghurst *et al.* 1998). Exposure to a mixture of non-ortho and mono-ortho-polychlorinated biphenyls had an adverse effect on reproduction in female *Fundulus heteroclitus* (Linnaeus) (Black *et al.* 1998). Development disturbances caused by polychlorinated biphenyls including impaired hatching, posterior malformations and oedemas and slow embryonic growth and anterior malformations have been found in the zebrafish (*Brachydanio rerio*) (Billson *et al.* 1998). Nonylphenol can effect testicular structure (degeneration of seminiferous lobules containing spermatocytes were found in the month of May and squamous Sertoli cells containing phagocytised sperm cells in the month of June) in the eelpout *Zoarces viviparus* (Christiansen *et al.* 1998).

A lack of oestrogenic synergy has been reported for mixtures of weakly oestrogenic hydroxylated polychlorinated biphenyls (OH-PCBs) and the pesticides endosulfan and dieldrin, even though all the individual OH-PCBs were oestrogenic in both the MCF-7 focus assay and a competitive oestrogen-receptor binding assay: of the pesticides only endosulfan was oestrogenic (Arcaro *et al.* 1998). In a range of oestrogen assays, certain binary mixtures of environmental oestrogens (pesticides) have been shown to act additively (but not synergistically) (Ramamoorthy *et al.* 1998).

Excreted endogenous and synthetic oestrogens

Natural oestrogens such as the 17β-oestradiol and oestrone released from human faeces, have been identified in sewage lagoon water fractionates where they are thought to be one of the main causes of observed hermaphroditism in male fish (Environment Agency Report 1996, 1998; Routledge *et al.* 1998; Desbrow *et al.* 1998; Wiseman *et al.* 1998a, b). Concentrations of both 17β-oestradiol and oestrone were as high as 48 ng/l in effluent from sewage-treatment works at one of the sites tested in Essex. In comparison, only small quantities of the synthetic oestrogen 17α-ethylnyloestradiol, derived from the contraceptive pill were detected, although 7 ng/l was detected on one occasion in effluent from the site: an oestrone concentration of 76 ng/l was also detected. The factors involved in the excretion of endogenous oestrogens require further research: consumption of ethanol appears to be one factor implicated in enhancing the excretion of 17β-oestradiol.

Newly hatched fry of the Japanese medaka (*Oryzias atipes*) exposed to 17β-oestradiol developed primarily into females or had testis–ova (Hartley *et al.*, 1998). Exposure of fathead minnows (*Pimephales promelas*) to 17β-oestradiol resulted in reproductive impairment (Kramer *et al.* 1998). In the rainbow trout (*Oncorhynchus mykiss*) exposure to 17β-oestradiol downregulates

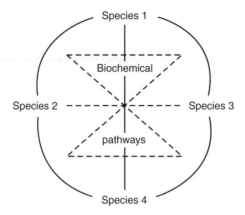

Figure 7.1. Networking webs in ecostasis

hepatic flavin-containing monooxygenase activity, expression of which has been shown to be directly correlated with salinity in euryhaline fish (Schlenk *et al.* 1997).

Excreted dietary phytoestrogens

Phytoestrogens occur in a number of foodstuffs derived from plant components and include the isoflavones genistein and daidzein, (present mostly as their glycosides in soya bean-derived foods) and the lignan precursors in whole grains: (Wiseman 1999, see Chapter 6). Phytoestrogens and their metabolites may be present in waterways as a result of transit through animals with run-off into the aquatic environment or from urinary and faecal excretion of such phytoestrogens and their metabolites into domestic effluents. Phytoestrogens are known to be rapidly metabolized and excreted from the body in urine, faeces and breast milk. Hydrolysis of the isoflavone glycoside conjugates genistin and daidzin, by bacterial β-glucosidase in the human gut, produces genistein and daidzein (more biologically active forms). The oestrogenic metabolite equol can be formed from daidzein (*O*-desmethylangolensin is also formed) by further bacterial metabolism and is then reabsorbed. The increased consumption of phytoestrogens as health foods could thus increase the amounts of phytoestrogens and their metabolites that are excreted and, hence their contribution to the amount of environmental oestrogenic material. This requires biomonitoring and possibly bioremediation (Lynch and Wiseman 1998).

Development of replacement materials for xenoestrogen avoidance

The UK has agreed to phase out the use of these nonylphenol ethoxylates as industrial cleaning agents by the year 2000 (Paris Convention). Any problem of phthalates leaching out from plastic food packaging could be solved by replacing plastic containers with aluminium foil, glass or greased paper (Ridgway and Wiseman 1998). Alternatively, new biodegradable plastics, particularly those based on plant sources, for example, oilseed rape (Slabas *et al.* 1993), could be designed so as not to leach oestrogenic substances. Such products with their designed susceptibility to biodegradation, would in addition present no long-term landfill problems. This could subsequently reduce the potential for oestrogenic substances to leach into ground water and the water supply. A general principle of design of new products to minimise oestrogenicity would be to avoid the use of *p*-hydroxyphenols, which readily match the molecular oestrogenicity template described above.

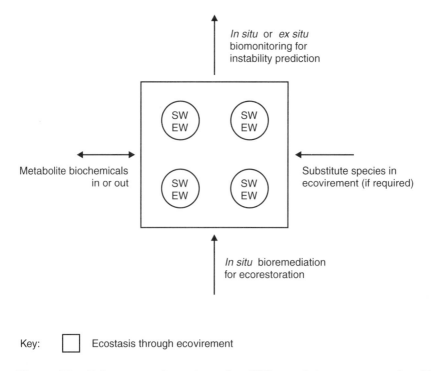

Figure 7.2. Robustness of species-webs (SW) overlying enzyme-webs (EW): achievement of ecostasis through ecovirement

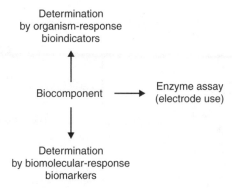

Figure 7.3. Environmental-biomonitoring: assay modes

Activated-carbon filtration of water in the removal of environmental oestrogens and oestrogen-mimics

In addition to the widespread interest in the use of activated carbon in cleaning air (it is a vital component in gas masks and filtration units) (Evans 1999), oestrogenic substances may be removed from water by the use of activated-carbon filters (Ridgway and Wiseman 1998). The large-scale production of high-quality activated-carbon filters, however, represents an environmental problem. Developments, such as the production of carbon-ised plant waste could therefore be of great value in the future. Waste plant sources include peanut shells and slash pine bark from fibre board production (Edgehill and Lu 1998). Plant material is carbonised, activation of quality carbon sources (wood or coal) is carried out to produce commercial activated carbons. Activation involves exposure to high-temperature oxidising gases, for example, steam or air which substantially increases the surface area for absorption by producing more micropores. Activation may, however, represent over-engineering particularly as phenols, including oestrogenic types, are susceptible to irreversible adsorption which prevents regeneration. The single use of less expensive carbonised substitute from waste plant material may therefore be more effective, economically and environmentally, than the employment of currently used carbon filters.

Use of charged suspended-particles to remove environmental oestrogens and oestrogen-mimics

Ion exchange may be of value in the removal of oestrogenic substances, especially in conjunction with chemical or enzymic modification which

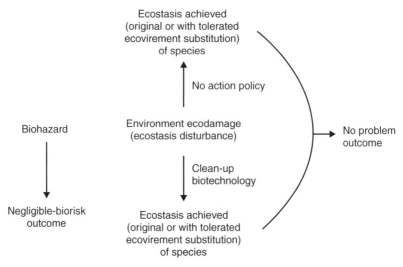

Figure 7.4. Environmental-bioremediation: for biorestoration or ecovirement

would increase oestrogen polarity (Ridgway and Wiseman 1998). The study of Sun *et al.* (1992) utilised chitosan to remove the oxidised (quinonic) and charged phenolics. Relatively cheap media, for example, chitosan would appear to be preferable, to high-cost synthetic ion exchange dextrans for large-scale environmental application. Coagulants such as polyethylene-amine are also similarly of value. Wada *et al.* (1995) used this approach to remove phenolics from solution using fungal tyrosinase, in this case immobilised on magnetite particles.

Enzymes may be used to produce charged oestrogen-binding materials from plant waste. Apple juice is clarified by a system which utilises polygalacturanase and pectin methyl esterases to produce aggregating charged protein-pectic globules (Lea 1995). Collected precipitate or specially processed waste apple could conceivably therefore be used as the basis for the development of an oestrogen/pollutant removal scheme based upon flocculation or other applications of colloid science.

Chemical catalysis for the removal of environmental oestrogens and oestrogen-mimics

Oxidation/hydroxylation of phenolic substances may be achieved by chemical catalysis (Ridgway and Wiseman 1998) including palladium catalysts. If this were combined with developments in the use of zeolites and or bio-membrane technology (Parton *et al.* 1994) systems for the removal of oestrogenic substances may be produced. The zeolite would act as a support and allow for the ready removal and regeneration of the catalyst. Embedding the zeolite in a synthetic membrane would provide a relatively non-polar surface suitable for oestrogen degradation. Additional alkane degradative catalytic activity could be provided by iron-based catalysts (iron phthalocyanine) which would also be expected to result in polymerisation and hence complete oestrogen inactivation. This, however, could make the system more difficult to regenerate for reuse.

Use of enzymic catalysis in the removal of environmental oestrogens and oestrogen-mimics

Most of the enzymes that could have the ability to chemically modify/ degrade particular oestrogens and oestrogen-mimics, depend upon the redox reactions of transition metals. These enzymes include cytochromes P-450 (with nearly 1000 isoforms possible), polyphenol oxidases, laccases, peroxidases and the di-oxygenase ring-splitting enzyme found in white rot fungi, including *ortho* (catechol 1,2 dioxygenases) and *meta* (catechol 2,3 dioxygenase) types. Novel types are also likely to be discovered (or constructed) in the future, for example, enzymes from brown algae are capable of hydroxylating progesteron (Pollio *et al.* 1996), although in this particular case, oestrogenicity would probably be increased rather than decreased, the principle of identifying sterol and oestrogen metabolising enzymes from novel sources is illustrated (appropriate enzyme-mimics and genetically or chemically modified enzymes are also being developed). Strategies for the enzymic degradation of oestrogenic substances include: (1) introduce further hydroxyl groups preferably in the *ortho* position to enable either (2) the formation of quinones which can be removed from solution by charge association or polymerisation, or (3) ring cleavage by *ortho* or *meta* dioxygenases. If (3) takes place within a living micro-organism the degradation products are subsequently metabolised as Krebs cycle intermediates. A good example of the use of an enzyme (fungal tyrosinase) to remove phenolics (*p*-cresol and 4-methyl catechol) from solution has been reported by Sun *et al.* (1992).

Phase II detoxification reactions as performed in the mammalian liver could be considered to be a useful model in devising future systems for oestrogenic removal. Group addition, for example, glucoronidation and sulphation, added to mask functionality or improve solubility may, however, conceivably be reversed in the environment. Phase II reactions therefore may be of limited value in oestrogen bioremediation without non-biodegradable ligand attachment to phenolic functional groups. If isolated enzymes are considered, the requirement for an expensive cofactor (NADPH) could potentially be limiting for cytochromes P-450. Cytochromes P-450 could be essential, however, for deactivating sterol oestrogens by hydroxylation. Cytochromes P-3450 enzymes are not widely used industrially because they have a poor turnover, low stability and need a regeneration system for the enzyme cofactor (Sutherland 1999). Recently, however, it has been possible to enhance the normally inefficient peroxide shunt pathway of one of the P-450 enzymes, thus making the need for a cofactor regeneration system redundant (Joo *et al.* 1999; Sheldon 1999; Sutherland 1999).

Polyphenol oxidase (see below) would seem to be at a distinct advantage in its potential to render phenolics harmless as it requires only molecular oxygen in addition to substrate. The problem with some polyphenol oxidases is that they possess little or no *p*-monophenolic oxidase capability. An apple polyphenol oxidase, however, has been shown to be highly capable of *p*-monophenolic oxidation. Furthermore, it should prove possible to produce it on a large scale at a relatively low cost (Ridgway and Wiseman 1998).

Removal of oestrogenic substances may also be possible by the large-scale use of monoclonal antibodies. Such antibodies are being prepared by a number of laboratories as part of the development of assays of oestrogenic substances, but could conceivably also have a role in oestrogen removal if they could be produced for low cost. In the future, oestrogen catalytic antibodies may be produced. One approach to this could be the production of fusion proteins consisting of oestrogen-binding epitopes and low molecular weight phenolic-oxidising protein fragments, such as the active 24 kDa fragments of apple polyphenol oxidase (Ridgway and Wiseman 1998). Other forms of affinity binding of oestrogens and oestrogen-mimics based upon receptors may prove to be useful in the future. Antibody and enzyme systems could be particularly successful in applications where the inactivation of small-scale high concentrations of oestrogens is practised. An example of this is the requirement to de-oestrogenise the spent patches used in hormone replacement therapy (HRT), where difficulty is currently experienced in ensuring the incineration of the patches.

Plant polyphenol oxidases: awaiting utilization

Polyphenol oxidases (Ridgway *et al.* 1997) are awaiting utilisation in the removal of environmental oestrogens and mimics (Ridgway and Wiseman 1998). Investigation of polyphenol oxidase reveals it to be plastid-associated (in particular with the thylakoid membrane and PSII: Vaughan *et al.* 1988) ubiquitous copper containing plant enzyme, which catalyses the hydro-xylation of monophenols to *p*-diphenols (monophenol monooxygenase or tyrosinase activity, EC 1.14.18.1) and the oxidation of *o*-diphenols to *o*-quinones utilising molecular oxygen (catechol oxidase or diphenol oxidase or diphenol oxygen oxidoreductase activity, EC 10.3.2). The laccases, found in fungi and higher plants (E.C.1.10.3.1) oxidise *p*-diphenols in addition to *o*-diphenols to their corresponding quinones (Mayer 1987). The organisation of the polyphenol oxidase gene family has been described for tomato (Newman *et al.* 1993). Polyphenol oxidase in potato (*Solanum tuberosum*) has been reported to be inducible by systemic wounding. Only those tissues which were developmentally competent to express polyphenol oxidase mRNA were capable of responding to the systemic wound signal by increased accumulation of polyphenol oxidase mRNA (Thipyapong *et al.* 1995). Moreover, an apple polyphenol oxidase cDNA has been shown to be upregulated in wounded tissues and this again suggests transcriptional control of polyphenol oxidase after wounding (Boss *et al.* 1995) and a potential role in defence mechanisms.

Ironically, polyphenol oxidase is generally considered to be a difficult enzyme to purify because of the presence of phenolics with which it reacts, resulting in the modification and inactivation of the protein molecule (Mayer 1987). In addition, solubilisation of particulate polyphenol oxidase is a difficult procedure. However, a number of methods for its extraction and purification from apple have been reported. Partial purification of polyphenol oxidase has been achieved (by ammonium sulphate precipita-tion and hydrophobic chromatography on Phenyl Sepharose CL4B) from the fruit of 12 cultivars grown in France, analysed for polyphenol oxidase activity in both the cortex and peel (Janovitz-Klapp *et al.* 1989). Enzyme activity ranged from 0.62 to 3.1 mkat/kg in the cortex and from 0.3–3 mkat/ kg in the peel, it is of interest that polyphenol oxidase activity was always equivalent or lower in the peel than in the cortex (Janovitz-Klapp *et al.* 1989). The optimum conditions for polyphenol oxidase extraction from apple fruit were achieved using a buffer with a pH > 7.0 and containing 15 mM ascorbic acid and 0.5% Triton X100. The Red Delicious apple cultivar showed the highest polyphenol oxidase activity and the Elstar cultivar the lowest and polyphenol oxidase has been purified 120-fold from the cortex of Red Delicious with a yield of around 40% (Janovitz-Klapp *et al.* 1989). The optimum pH for maximum activity 4.5–5 for the substrates methylcatechol, chlorogenic acid and (+)catechin and the K_m values were around 5mM for the three substrates and were independent of pH on the acid side of the pH optimum (Janovitz-Klapp *et al.* 1989). Chlorogenic acid was a better substrate for this apple polyphenol oxidase than catechin at pH 4, which is closer to the natural pH of apple vacuoles (Janovitz-Klapp *et al.* 1989). It is of interest that the polyphenol oxidase activity declined steadily in the Red Delicious apples in the eight weeks leading up to the commercial harvest date, resulting in an overall decrease in enzyme activity of around 30% (Janovitz-Klapp *et al.* 1989).

Moreover, partial purification has now been achieved for the Amaysa apple by ammonium sulphate precipitation and dialysis, which has one of the highest rates of enzymatic browning among several apple cultivars; indicating high levels of polyphenol oxidase (Oktay *et al.* 1995). The optimum pH for the substrates catechol, 4-methyl catechol, pyrogallol and L-dopa were 7.0, 9.0, 8.6 and 6.6 respectively. Catechol was found to be the most suitable substrate for Amaysa apple polyphenol oxidase, and 18 °C was the optimum temperature for maximum polyphenol oxidase activity with catechol as substrate. Separation by electrophoresis enabled the detection of three isoenzymes with catechol and L-dopa substrates. In relation to control of browning a number of inhibitors was tested, and their order of effectiveness was L-cysteine > sodium metabisulphite > ascorbic acid > sodium cyanide > mercaptoethanol > glutathione > thiourea (Oktay *et al.* 1995).

Polyphenol oxidase has been extensively purified from apple flesh (*Malus pumila* cv. Fuji); it was purified 470-fold from the plastid fraction by ammonium sulphate precipitation, gel filtration and ion-exchange chromatography with a total yield of around 70% (Murata *et al.* 1992). The M_r was found to be around 65 000 by both gel filtration chromatography and sodium dodecyl sulphate-polyacrylamide gel electrophoresis, and the N-terminal amino acid sequence was N-Asp-Pro-Leu-Ala-Pro-Pro (Murata *et al.* 1992). The optimum pH for enzyme activity was around pH 4 and the enzyme was stable in the pH range 6–8 (Murata *et al.* 1992). The K_m of the enzyme for the substrate, chlorogenic acid, was 0.122 mM and indeed the rate of reaction of the purified enzyme was much greater for chlorogenic acid than for other p-diphenols such as (+)catechin, (−)epicatechin and 4-methylcatechol. In addition, the enzyme lacked both monophenol monooxygenase and p-diphenol oxidase activity (Murata *et al.* 1992). The purified enzyme was found to be much less thermally stable than the enzyme of the plastid fraction (Murata *et al.* 1992).

Apples (*Pyrus malus* L. cv. Granny Smith) have been used in the purification of an active proteolysed isoform of apple pulp polyphenol oxidase achieved by a rapid three-step method based on the resistance of polyphenol oxidase to further sodium dodecyl sulphate-proteinase K digestion (Marques *et al.* 1994). Extraction from the thylakoid membrane pellet and pre-purification by temperature-induced phase partitioning, was followed by sodium dodecyl sulphate-proteinase K digestion and then purification to 388-fold homogeneity by DEAE-cellulose column chromatography (Marques *et al.* 1994). A yield of >40% was achieved and this active polyphenol oxidase isoform was used to raise polyclonal antibodies, resulting in the production of highly titred specific serum used to perform immunoblots to detect active and latent forms of the enzyme (Marques *et al.* 1994). The sodium dodecyl sulphate-proteinase K digestion had no effect on the apparent K_m at pH 4.6, which was 7 mM for 4-methylcatechol before digestion and 6.7 mM after (Marques *et al.* 1994). These results are comparable to those obtained with polyphenol oxidase from Red Delicious apples, at the same pH (Janovitz-Klapp *et al.* 1989).

Following the isolation and purification of apple-derived polyphenol oxidase, considerable effort has been put into the purification of potato tuber (*Solanum tuberosum* cv. Cara) polyphenol oxidase free from the storage protein patatin (Partington and Bolwell 1996). In potato, patatin is the major storage protein in the tuber and often contaminates preparations. Purification of polyphenol oxidase from the potato tuber has been achieved using the important step of hydrophobic chromatography on Octyl–Sepharose to completely remove patatin (Partington and Bolwell 1996). The resulting purified polyphenol oxidase had a K_m of 4.3 ± 0.3 mM for L-dihydroxyphenylalanine and was shown to be a doublet of M_r 60 000 and 69 000 when analysed by sodium dodecyl sulphate-polyacrylamide gel electrophoresis (both bands had similar N-termini corresponding to polyphenol oxidase isoforms when sequenced) (Partington and Bolwell 1996). Purification of polyphenol oxidase from carrot (M_r 59 000) has also been achieved (Soderhal 1995).

In healthy leaves, polyphenol oxidase is bound to the thylakoid membranes of the chloroplast, but it is not an intrinsic membrane protein and can be released from the thylakoids by sonication, mild detergent treatment or protease treatment. Anionic detergents such as sodium dodecyl sulphate and proteases such as trypsin can activate the latent activity of polyphenol oxidase, and the effect of sodium dodecyl sulphate on polyphenol oxidase (from broad bean leaf *Vicia faba*) has been investigated (Jimenez and Garcia-Carmona 1996). This particular polyphenol oxidase was enzymatically inactive in aqueous buffers at neutral pH and active at acid pH (pH 3–4); however, in the presence of sodium dodecyl sulphate the activity at acid pH was eliminated and the monophenol monooxygenase and catechol oxidase activities at neutral pH were activated (Jimenez and Garcia-Carmona 1996). This rapid activation was dependent on the concentration of sodium dodecyl sulphate used. The relationship demonstrated between sodium dodecyl sulphate concentrations and proton concentrations may be related to a displacement of the sensitive pKs of the enzyme by interaction with sodium dodecyl sulphate molecules (Jimenez and Garcia-Carmona 1996). Furthermore a polyphenol oxidase partially purified from broad bean seeds has been shown to oxidise the flavonol fisetin, see Chapter 1 (Jimenez *et al.* 1998).

Alternative, potentially economic methods have been developed for the purification of apple polyphenol oxidase (Ridgway and Tucker 1999). The yields of this polyphenol oxidase were high enough such that alternative potential production methods, such as expression of the enzyme in micro-organisms, would not be a viable economic alternative. This apple polyphenol oxidase was used to produce oxidation products of phloridzin including the potent antioxidant 3-hydroxyphloretin (see Chapter 1).

Microbial bioremediation of environmental oestrogens and oestrogen-mimics

Use of chemical catalysts and enzymes to remove oestrogens from the environment and food has been considered (see above). Most bioremedi-ation, however, is carried out by whole organisms (Ridgway and Wiseman 1998). At sewage-treatment plants, micro-organisms are provided with an aerated environment in gravel beds, or other bio-reactors (activated sludge process). Purification of water effluents is accomplished therefore by assimilation by micro-organisms of the organic matter in this sewage and its resynthesis into the living organisms which aggregate as flocs in a gelatinous matrix. This process thus changes organic matter from colloidal (and dissolved) states of dispersion to a state in which it will settle out on to the bed, leaving the purified water for recycling. It is particularly important that this process is efficient in urban areas, because water is recycled domestically about five times and therefore has potential for significant accumulation of oestrogens. This is particularly so in lagoons where protein-bound oestrogens may settle out: 17β-oestradiol, which is normally protein-bound in human faeces, can be released during the waste treatment process. To avoid this oestrogen leakage improving the binding properties of the flox matrix and the particular inoculation and mix of organisms used is a necessary goal. Aerobic conditions are required for the degradative/ metabolic processes of oestrogen remediation described above. It has recently been suggested, however, that the lack of benzene degradation in the sulphate reduction zone of some aquifers may result from the failure of the appropriate benzene degrading sulphate reducers to colonise the aquifers rather than from adverse environmental conditions. This may have implications for the accumulation of oestrogens in largely anaerobic lagoons and the possibility of inoculating them to achieve the biodegrada-tion of oestrogens (Weiner and Lovely 1998).

There is potential to genetically engineer organisms to readily degrade environmental pollutants (Timmis and Pieper 1999) including oestrogenic substances. Genetic engineering incorporating non-endogenous cyto-chromes P-450 could be particularly useful for the metabolism of sterols.

It would be important, however, to take into account the need to co-express the cytochrome P-450 reductase in addition to the required cytochrome P-450 activity. This general principle follows studies by van den Brink *et al.* (1996), where the introduction of multiple copies into *Aspergillus niger* of the *A. niger bph* A gene, which encodes the cytochrome P-450 enzyme benzoate *P*-hydroxylase, did not result in increased activities of this enzyme. Increased levels of enzyme activity were only observed following the co-expression of multiple copies of the *A. niger cpr* A gene which encoded cytochrome P-450 reductase. An example of whole pathway engineering for the metabolism of aromatic compounds in general is provided by Panke *et al.* who genetically engineered strains of *Pseudomonas putida* (using the upper TOL operon of plasmid pWW0) to grow on toluene as the sole carbon source (Panke *et al.* 1998). The organisms were engineered using Tn-5 derived transposon vectors, which have since 1990 allowed for stable chromosomal insertion. The use of excisable selection markers was used as the presence of antibiotic resistance genes in organisms destined for environmental release is undesirable (Panke *et al.* 1998). In considering the transformation of micro-organisms it is important to consider the target site for expression, for example, exported proteins could be of particular value in removing oestrogens from the aquatic environment. Many problems may also be encountered when trying to express non-endogenous proteins in fungi. General principles to help expression include: the introduction of introns to improve mRNA stability and the elimination of AT-rich regions as appropriate, the improvement of mRNA stability by fusion with highly expressed genes, the co-expression foldases and chaperones, and the use of protease-deficient strains (Gouka *et al.* 1997).

The use of plants for bioremediation of environmental oestrogens and oestrogen-mimics

Plants have great potential for bioremediation (Ridgway and Wiseman 1998). The most successful example of plant-based bioremediation is the application of reed bed technology to treat the effluent of chemical plants (Davies *et al.* 1994). Reed beds, as part of reconstituted wetlands, not only have the potential to improve the quality of water supply but also the potential to increase the diversity of wildlife in a particular area. The urine and faeces of farm livestock (female or oestrogen-supplemented) present a potential source of oestrogens (and their metabolites) in the environment. Grasses transformed to metabolise oestrogens could perhaps help to reduce the oestrogen content of water runoff. Plants transformed to degrade oestrogens, particularly if used in monoculture beds, may be better than transformed micro-organisms when used in sewage farms, because these farms utilise mixed populations. Here oestrogens would be minority substrates, and the expression of the oestrogen metabolising genes would represent an additional burden in a competitive environment.

A general consideration when using genetic engineering to facilitate the generation of bioremediating organisms is the resistance of the organism to pollution-induced stress in relation to the likely pollutant stress to be encountered. With exposure to high levels of pollutants it may be imperative to co-transform organisms with, in addition to cytochrome P-450 reductase genes, etc., general stress resistance genes, for example *gor*, *gsh* I and II, coding for glutathione reductase, γ-glutamylcysteine synthetase and glutathione synthetase respectively (Foyer *et al.* 1994). This again, however, would reduce the efficiency of use of primary metabolites, and prove uncompetitive in multi-organism, largely non-toxic environments. An advantage of plants over most micro-organisms for the generation of genetically engineered bioremediative capability is that transformation of plastid genes would enable containment of non-endogenous genes (plastid transgenes are not transmitted by pollen) within the engineered plant and thus prevent gene flow to the environment at large (Daniell *et al.* 1998).

A healthy biosphere: progress towards 'Techno-Gaia'

The oestrogen receptor is a member of a large family of receptors, and other members are also likely to be affected by micro-organic-pollutants, perhaps the most potent of which could be thyroxine mimics. There is also concern at least in the natural environment over female reproductive health, for example the use of the maritime antifouling agent, tributyltin, giving rise to the condition of apposex in molluscs. Further strategies for environmental oestrogen removal, in addition to those discussed above, clearly need to be considered, for example a recent report has shown that titanium dioxide activated with UV light can convert oestrogen to carbon dioxide, and the removal of oestrogen from water by this novel method is now being tested in a pilot plant. There is also great potential for exploiting antibody-based technologies to manage environmental pollution (Harris 1999). Indeed, the production of antibodies to oestrogens, in addition to giving opportunities for oestrogen removal, also gives the opportunity to design biosensors for oestrogenic substances. For example, direct piezoelectric immunosensors for use in solution have been constructed for the herbicide, atrazine (Steegborn and Skladal 1996). Remote monitoring for the detection of oestrogenic pollutants in the environment is thus clearly feasible. Such detection systems could be designed to release enzymes/inoculums to degrade oestrogenic substances in addition to employing automated reporting to monitoring centres. Perhaps this 'Techno-Gaia' is a vision of the future; it would incorporate and accommodate the natural world where possible, but would also utilise ecomonitoring with fail-safe and restorative systems (Wiseman and Lynch 1998). With the increasing adoption worldwide of the principle of 'polluter pays', for example in the UK Environment Act of 1995, and the commercial awareness that a profit may be made from environmental protection, this clean future seems increasingly likely.

Biomonitoring of environmental oestrogens and oestrogen-mimics: leading to ecorestoration?

Novel, rapid and inexpensive bioassessment techniques need to be developed to facilitate the identification of oestrogen-mimics (full or partial agonists or indeed antagonists) by biomonitoring, and subsequent bioremediation where necessary of aquatic environments (Wiseman *et al.* 1998a, b). For total human exposure to oestrogenic material and for water-environment ecology and ecotoxicology, environmental oestrogen effluent derivatives (xenoestrogens) may be of less significance than excreted endogenous oestrogens and excreted dietary phytoestrogens in sewage and from land runoff. Assessing the extent of active oestrogenic material in food is a related and important issue. In order to predict, which compounds are likely to be active as endocrine disruptors in water or soil, the new approaches that need to be developed should utilise the ability of oestrogenic compounds to bind to membranes, as well as to be regulators of gene expression.

Prediction of oestrogen-mimicry by biomolecular modelling techniques

The potential of environmental contaminants for oestrogen mimicry will necessitate the use of a variety of specific laboratory tests. In the case of agonism or antagonism of the ER, these could include the production of the yolk protein vitellogenin (an oestrogen-dependent process) in male fish or the use of engineered yeast cells expressing human ER (Environment Agency Report, 1996, 1998; MRC/IEH report, 1995). Indeed, the endocrine modulating actions of β-sitosterol have been evaluated using a series of *in vitro* and *in vivo* assays: all three assays, i.e. receptor binding assay, vitellogenin production in hepatocytes and *in vivo* vitellogenin production in sexually immature trout, all confirmed the oestrogenicity of β-sitosterol in trout (Tremblay and van der Kraak 1998). The *in vivo* synthesis of vitellogenin has been used as a biomarker for xenoestrogens, including nonylphenol, bisphenol A and dibutylphthalate in the rainbow trout (*Oncorhynchus mykiss*) (Christiansen *et al.* 1998) and nonylphenol can induce vitellogenin in the eelpout (*Zoarces viviparus*) (Christiansen *et al.* 1998). The sensitivity of fish to exposure to oestrogenic contaminants is species dependent and the Atlantic cod has been proposed as a suitable marine species for monitoring the effect of environmental oestrogens (by measurement of plasma vitellogenin) (Hylland and Haux 1997).

Most studies of oestrogen-mimics in fish have focused on the stimulation of vitellogenin synthesis by the liver, it is of interest then that p, p'-DDT has been shown to act on the hypothalmic–pituitary–gonadal axis to stimulate gonadotropin release in the Atlantic croaker, a teleost model (Khan and Thomas 1998). Exposure of fathead minnows (*Pimephales promelas*) to 17β-oestradiol resulted in induction of alkaline-labile phosphate, a biomarker of oestrogen exposure (Kramer *et al.* 1998). A recombinant yeast system expressing the trout oestrogen receptor and trout hepatocyte cultures are two bioassays that have been used successfully for screening the oestrogenic potency of oestrogen-mimics (Petit *et al.* 1997).

The structural features of alkylphenolic chemicals associated with oestrogenic activity has been investigated in a yeast test system (expressing the human ER) and it was found that both the position (*para* > *meta* > *ortho*) and branching (tertiary > secondary=normal) of the alkyl group affect oestrogenicity: optimal oestrogenic activity required a single tertiary branched alkyl group composed of between six and eight carbon atoms located at the *para* position on a phenol ring that was otherwise unhindered (Routledge and Sumpter 1997). Bisphenol A has been shown to interact with the ER-α in a distinct manner from 17β-oestradiol: the pattern of activity of bisphenol A with ER-α mutants (inactivation of the AF-1 or AF-2 regions) different from that observed with weak oestrogens (oestrone and oestriol), partial ER-α agonists(4-hydroxytamoxigen and raloxifene) or the pure antagonist ICI 182, 780 indicating that bisphenol A is not just a weak oestrogen mimic but displays a distinct mechanism of action at the ER-α (Gould *et al.* 1998).

In terms of non-ER recognition sites, the importance of cell membranes in the action of endogenous oestrogens and oestrogen-mimics is now increasingly recognised. The liposomal membrane system is a well-tested assay for investigating of membrane lipid peroxidation and the action of membrane protective antioxidant compounds. Some antioxidants such as vitamin E protect the membrane via a chain-breaking action on the alkoxyl and peroxyl radical chain-reaction. 17β-oestradiol, tamoxifen and iso-flavone phytoestrogens also appear to act as membrane antioxidants, at least in part by entering the membrane and modulating membrane fluidity thus increasing membrane stability against lipid peroxidation (Wiseman *et al.* 1998c, see Chapter 1).

Environmental-contaminant oestrogen-mimics are also recognised by the liposome system. Bisphenol A displays a highly effective membrane antioxidant ability and inhibits liposomal-membrane lipid peroxidation in a concentration-dependent manner (Leadley *et al.* 1998). Similarly, a mixture of 4-nonylphenols (straight chain normal and isoforms) display a highly effective membrane antioxidant ability. These xenoestrogens mimic, therefore, not only the ER binding ability of 17β-oestradiol, but also its membrane antioxidant ability. Bisphenol A and 4-nonylphenol may also act by modulating membrane fluidity (Leadley *et al.* 1998). In addition to laboratory-based assays, determination by computer-assisted methods of structural similarity to 17β-oestradiol may prove to be a rapid test for environmental xenobiotics that might display xenoestrogen ability in water or soil.

Conclusions on ecorestoration after biomonitoring of environmental oestrogens and oestrogen-mimics

Understanding the relationship of biomolecular structure to oestrogen-mimicry (agonist or antagonist) is a *sine qua non* for prediction of hazards and benefits from environmental and dietary chemicals of whatever origin and structure. Dose-related 'cause and effect' must be clearly elucidated before reliable conclusions can be reached. Removal or even addition of oestrogen-mimics may become commonplace, dependent on reliable biomonitoring of locations. Rapid *in vitro* tests may compete with computer molecular modelling for reliability of mimicry prediction, but such prediction may take several (contradictory) forms in different biological environments, requiring a better understanding of multi-ligand binding sites.

In this context a transcription-independent signalling pathway for progesterone mediated by a membrane-bound receptor has recently been reported (Grazzini *et al.* 1998). Clearly, many steroids including the oestrogens and their environmental mimics can act at a variety of sites within the cell not, as had been thought previously, just with their transcription-modulating nuclear receptors. The role of the environmental oestrogens as either agonists or antagonists of normal hormonal regulation and as modulators of cellular damage will therefore require substantial reassessment.

Although ecorestoration by biointervention may be an achievable goal, ecovirement (substitution in species webs) may be an initial aim in many cases of environmental contamination by oestrogens and oestrogen mimics.

References

Arcaro, K.F., Vakharia, D.D., Yang, Y. and Gierthy, J.F. (1998) Lack of synergy by mixtures of weakly estrogenic hydroxylated polychlorinated biphenyls and pesticides. *Environ. Health Perspect.*, **106**(Suppl. 4), 1041–1046.

Billinghurst, Z., Clare, A.S., Fileman, T., McEvoy, J., Readman, J. and Depledge, M.H. (1998) Inhibition of barnacle settlement by the environmental oestrogen 4-nonylphenol and the natural oestrogen 17β-oestradiol. *Mar. Pollut. Bull.*, **36**, 833–839.

Billson, K., Westerlund, L., Tyskind, M. and Olsson, P.-E. (1998) Developmental disturbances caused by polychlorinated biphenyls in zebrafish (*Brachydanio rerio*). *Mar. Environ. Res.*, **46**, 461–464.

Black, D.E., Gutjahr-Gobell, R., Pruell, R.J., Bergen, B. and McElroy, A.E. (1998) Effects of a mixture of non-ortho- and mono-ortho-polychlorinated biphenyls on reproduction in *Fundulus heteroclitus* (Linnaeus). *Environ. Toxicol. and Chem.*, **17**, 1396–1404

Boss, P.K., Gardner, R.C., Janssen, B.-J. and Ross, G.S. (1995) An apple polyphenol oxidase cDNA is up-regulated in wounded tissue. *Plant Molec. Biol.*, **27**, 429–433.

Brzozowski, A.M., Pike, A.C.W., Dauter, Z., Hubbard, R.E., Bonn, T., Engstrom, O., Ohman, L., Greene, G.L., Gustafsson, J.-A. and Carlquist, M. (1997) Molecular basis of agonism and antagonism in the oestrogen receptor. *Nature* **389**, 753–758.

Christiansen, T., Korsgaard, B. and Jespersen, A. (1998) Induction of vitellogenin synthesis by nonylphenol and 17β-estradiol and effects on the testicular structure in the eelpout *Zoarces viviparus*. *Mar. Environ. Res.*, **46**, 141–144.

Christiansen, L.B., Pedersen, K.L., Korsgaard, B. and Bjerregaard, P. (1998) Estrogenicity of xenobiotics in rainbow trout (*Oncorhynchus mykiss*) using in vivo synthesis of vitellogenin as a biomarker. *Mar. Environ. Res.*, **46**, 137–140.

Daniell, H., Datta, R., Varma, S., Gray, S. and Lee, S.-B. (1998) Containment of herbicide resistance through genetic engineering of the chloroplast genome. *Nature Biotechnol.*, **16**, 345–348.

Desbrow, C., Routledge, E.J., Brighty, G.C., Sumpter, J.P. and Waldock, M. (1998) Identification of estrogenic chemicals in STW effluent. I. Chemical fractionation and in vitro biological screening. *Environ. Sci. and Technol.*, **32**, 1549–1558.

Edgehill, R.U. and Lu, G.Q. (1998) Adsorption characteristics of carbonized bark for phenol and pentachlorophenol. *J. Chem. Technol. Biotechnol.*, **71**, 27–34.

Environment Agency Report (1996) *The Identification and Assessment of Oestrogenic Substances in Sewage Treatment Effluents.* UK Stationery Office, London.

Environment Agency Report (1998) *Endocrine-disrupting Substances in the Environment: What Should be Done?* UK Stationery Office, London.

Evans, A. (1999) Cleaning air with carbon. *Chem. and Ind.*, **18**, 702–795.

Gibbons, W.N., Munkitirick, K.R., McMaster, M.E. and Taylor, W.D. (1998) Monitoring aquatic environments receiving industrial effluents using fish species 2: Comparison between responses of trout-perch (*Percopsis omiscomaycus*) and white sucker (*Catostomus commersoni*) downstream of a pulp mill. *Environ. Toxicol. and Chem.*, **17**, 2238–2245.

Gimeno, S., Komen, H., Gerritsen, A.G.M. and Bowmer, T. (1998) Feminisation of young males of the common carp, *Cyprinus carpio*, exposed to 4-tert-pentylphenol during sexual differentiation. *Aquatic Toxicol.*, **43**, 77–92.

Gimeno, S., Komen H., Jobling, S., Sumpter, J. and Bowmer, T. (1998a) Demasculinisation of sexually mature male common carp, *Cyprinus carpio*, exposed to 4-tert-pentylphenol during spermatogenesis. *Aquatic Toxicol.*, **43**, 93–109.

Gould, J.C., Leonard, L.S., Maness, S.C., Wagner, B.L., Conner, K., Zacharewski T., Safe, S., McDonnell, D.P. and Gaido, K.W. (1998) Bisphenol A interacts with the estrogen receptor α in a distinct manner from estradiol. *Molec. and Cell. Endocrinol.*, **142**, 203–214.

Gouka, R.J., Punt, P.J. and van den Hondel, C.A.M.J.J. (1997) Efficient production of secreted proteins by *Aspergillus*: progress, limitations and prospects. *Appl. Microbiol. Biotechnol.*, **47**, 1–11.

Harris, B. (1999) Exploiting antibody-based technologies to manage environmental pollution. *Trends in Biotechnol.*, **17**, 290–296.

Hartley, W.R., Thiyagarajah, A., Anderson, M.B., Broxson, M.W., Major, S.E. and Zell, S.I. (1998) Gonadal development in Japanese medaka (*Oryzias latipes*) exposed to 17β-estradiol. *Mar. Environ. Res.*, **46**, 145–148.

Hyland, K. and Haux, C. (1997) Effects of environmental oestrogens on marine fish species. *TRAC – Trends in Analyt. Chem.*, **16**, 606–612.

Janovitz-Klapp, A., Richard, F. and Nicolas, J. (1989) Polyphenoloxidase from apple, partial purification and some properties. *Phytochemistry*, **28**, 2903–2907.

Jimenez, B. (1997) Environmental effects of endocrine disruptors and current methodologies for assessing wildlife health effects. *TRAC – Trends in Analyt. Chem.*, **16**, 596–606.

Jimenez, M. and Garcia-Carmona, F. (1996) The effect of sodium docecyl sulphate on polyphenol oxidase. *Phytochemistry*, **42**, 1503–1509.

Joo, H., Lin, Z. and Arnold, F.H. (1999) Laboratory evolution of peroxide-mediated cytochrome P450 hydroxylation. *Nature*, **399**, 670–673.

Katzenellenbogen, J.A., O'Malley, B.W. and Katzenellenbogen, B.S. (1996) Tripartite steroid hormone receptor pharmacology: interaction with multiple effector sites as a basis for the cell- and promoter-specific action of these hormones. *Mol. Endocrinol.*, **10**, 119–131.

Khan, I.A., Thomas, P. (1998) Estradiol-17β and o,p'-DDT stimulate gonadotropin release in Atlantic croaker. *Mar. Environ. Res.*, **46**, 149–152.

Kramer, V.J., Miles-Richardson, S., Pierens, S.L. and Giesy, J.P. (1998) Reproductive impairment and induction of alkaline-labile phosphate, a biomarker of estrogen exposure, in fathead minnows (*Pimephales promelas*) exposed to waterborne 17β-estradiol. *Aquatic Toxicol.*, **40**, 335–360.

Lea, A.G.H. (1995) Enzymes in the production of beverages and fruit juices. In *Enzymes in Food Processing*, Tucker, G.A. and Woods, L.F.J. eds. Blackie, London, pp. 223–247.

Leadley, J., Lewis, D.F.V., Wiseman, H., Goldfarb, P.S. and Wiseman, A. (1998) Environmental oestrogen-mimics display membrane-antioxidant ability: importance of molecular modelling predictions. *J. Chem. Technol. Biotechnol.*, **73**, 131–136.

Marques, L., Fleuriet, A., Cleyet-Marel, J.-C. and Macheix, J.-J. (1994) Purification of an apple polyphenoloxidase isoform resistant to SDS-proteinase K digestion. *Phytochem.*, **35**, 1117–1121.

Mayer, A.M. (1987) Polyphenol oxidases in plants – recent progress. *Phytochem.*, **26**, 11–20.

Miksicek, R.J. (1995) *Environmental Oestrogens: Consequences to Human Health and Wildlife*. Institute for Environment and Health, Leicester.

MRC/IEH (1995) *Environmental Oestrogens: Consequences to Human Health and Wildlife*. Institute for Environment and Health, Leicester.

Murata, M., Kurokami, C. and Homma, S. (1992). Purification and some properties of chlorogenic acid oxidase from apple (*Malus pumila*). *Biosci. Biotech. Biochem.*, **56**, 1705–1710.

Newman, S.M., Eannetta, N.T., Yu, H., Prince, J.P., De Vicente, M.C., Tanksley, S.D. and Steffens, J.C. (1993) Organisation of the tomato polyphenol oxidase gene family. *Plant Molec. Biol.*, **21**, 1035–1051.

Oktay, M., Kufrevioglu, I., Kocacaliskan, I. and Sakiroglum, H. (1995) Polyphenoloxidase from Amasya apple. *J. Food Sci.*, **60**, 494–496.

Olea, N., Pazos, P. and Exposito, J. (1998) Inadvertent exposure to xenoestrogens. *Eur. J. Cancer Prev.*, **7**(Suppl. 1), S17–S23.

Panke, S., Sanchez-Romero, J.M. and de Lorenzo, V. (1998) Engineering of quasi-natural *Pseudomonas putida* strains for toluene metabolism through an *ortho*-cleavage degradation pathway. *Appl. Environ. Microbiol.*, **64**, 748–751.

Partington, J.C. and Bolwell, G.P. (1996) Purification of polyphenol oxidase free of the storage protein patatin from potato tuber. *Phytochemistry*, **42**, 1499–1502.

Parton, R.F., Vankelecom, I.F.J., Casselman, M.J.A., Bezoukhanova, C.P., Uytterhoeven, J.B. and Jacobs, P.A. (1994). An efficient mimic of cytochrome P-450 from a zeolite-encaged ion complex in a polymer membrane. *Nature*, **370**, 541–544.

Pazos, P., Olea-Serrano, M.F., Zuluaga, A. and Olea, N. (1998) Endocrine disrupting chemicals: Xenoestrogens. *Med. Biolog. Environ.*, **26**, 41–47.

Petit, K., LeGoff, P., Cravedi, J.-P., Valotaire, Y. (1997) Two complementary bioassays for screening the estrogenic potency of xenobiotics: Recombinant yeast for trout estrogen receptor and trout hepatocyte cultures. *J. Molec. Endocrinol.*, **19**, 321–335.

Pollio, A., Pinto, G., Greca, M.D., Fiorentino, A. and Previtera, L. (1996) Biotransformations of progesterone by *Chlorella* spp. *Phytochemistry*, **42**, 1499–1502.

Ramoorthy, K., Wang, F., Chen, I.-C., Safe, S., Norris, J.D., McDonnell, D.P., Gaido, K.W., Bocchinfuso, W.P., Korach, K.S., Ashby, J., Lefevre, P.A., Odum, J., Harris, C.A., Routledge, E.J. and Sumpter, J.P. (1998) Potency of combined estrogenic pesticides: Synergy between synthetic oestrogens? *Chemtracts*, **11**, 306–308.

Ridgway, T.J. and Wiseman, H. (1998) Removal of oestrogens and oestrogen mimics from the environment. *Biochem. Soc. Trans.*, **26**, 675–680.

Rivas, A., Olea, N., Olea-Serrano, F. (1997) Human exposure to endocrine-disrupting chemicals: Assessing the total estrogenic xenobiotic burden. *TRAC – Trends in Analyt. Chem.*, **16**, 613–619.

Routledge, E.J., Sheahan, D., Desbrow, C., Brighty, G.C., Waldock, M. and Sumpter, J.P. (1998) Identification of estrogenic chemicals in STW effluent. 2. In vivo response in trout and roach. *Environ. Sci. and Technol.*, **32**, 1559–1565.

Routledge, E.J. and Sumpter, J.P. (1997) Structural features of alkylphenolic chemicals associated with estrogenic activity. *J. Biol. Chem.*, **272**, 3280–3288.

Safe, S.H. (1998) Hazard and risk assessment of chemical mixtures using the toxic equivalency factor approach. *Environ. Health Perspect.*, **106** (Suppl. 4), 1051–1058.

Schlenk, D., E-Alfy, A. and Buhler, D.R. (1998) Down regulation of hepatic flavin-containing monooxygenase activity by 17β-estradiol in rainbow trout (*Oncorhynchus mykiss*). *Comp. Biochem. and Physiol. – Comp. Pharmacol. and Toxicol.*, **118**, 199–202.

Sheldon, R.A. (1999) Picking a winner. *Nature*, **399**, 636–637.

Soderhal, I. (1995) Properties of carrot polyphenol oxidase. *Phytochemistry*, **39**, 33–38.

Steegborn, C. and Skládal, P. (1996) Construction and characterization of the direct piezoelectric immunosensor for atrazine operating in solution. *Biosensors and Bioelectronics*, **12**, 19–27.

Sun, W.-Q., Payne, G.F., Moas, M.S.G.L., Chu, G.H. and Wallace, K.K. (1992) Tyrosinase reaction/chitosan adsorption for removing phenols from wastewater. *Biotechnol. Prog.*, **8**, 179–186.

Sutherland, J. (1999) Enzyme evolution. *Chem. and Ind.*, **19**, 745–748.

Thipyapong, P., Hunt, M.D. and Steffens, J.C. (1995) Systemic wound induction of potato (*Solanum tuberosum*) polyphenol oxidase. *Phytochemistry*, **40**, 673–676.

Timmis, K.N. and Pieper, D.H. (1999) Bacteria designed for bioremediation. *Trends in Biotechnol.*, **17**, 201–204.

Tremblay, L. and van der Kraak, G. (1998) Use of a series of homologous in vitro and in vivo assays to evaluate the endocrine modulating actions of beta-sitosterol in rainbow trout. *Aquatic Toxicol.*, **43**, 149–162.

Van den Brink, J.M., van den Hondel, C.A.M.J.J. and van Gorcom, R.F.M. (1996) Optimization of the benzoate-inducible *p*-hydroxylase cytochrome *P*450 enzyme system in *Aspergillus niger*. *Appl. Microbiol. Biotechnol.*, **46**, 360–364.

Vaughan, K.C., Lax, A.R. and Duke, S.O. (1988) Polyphenol oxidase: the chloroplast oxidase with no established function. *Physiol. Plant.*, **72**, 659–665.

Wada, S., Ichikawa, H. and Tatsumi, K. (1995) Removal of phenols and aromatic amines from wastewater by a combination treatment with tyrosinase and a coagulant. *Biotechnol. Bioeng.*, **45**, 304–309.

Weiner, J.M. and Lovely, D.R. (1998) Anaerobic benzene degradation in petroleum-contaminated aquifer sediments after inoculation with a benzene-oxidizing enrichment. *Appl. and Environ. Microbiol*, **64**, 775–778.

Wiseman, A., Goldfarb, P., Ridgway, T. and Wiseman, H. (1998a) Gender hazards of oestrogens and mimics in water environments. *J. Chem. Technol. Biotechnol.*, **71**, 3–5.

Wiseman, A., Goldfarb, P., Ridgway, T. and Wiseman, H. (1998b) Biomolecular site-recognition in the prediction of environmental mimicry. *Biochem. Soc. Trans.*, **26**, 670–674.

Wiseman, A. and Lynch, J.M. (1998) Development of 'fail-safe strategies' (FSS) to combat biomonitored biohazards from environmental xenobiotics. In: *Environmental Biomonitoring: the Biotechnology Ecotoxicology Interface*, J.M. Lynch and A. Wiseman, eds. Cambridge University Press, Cambridge, pp. 287–292.

Wiseman, H., O'Reilly, J., Lim, P, Garnett, A.P., Huang, W.-C. and Sanders, T.A.B.S. (1998) Antioxidant properties of the isoflavone phytoestrogen functional ingredient in soya products. In: *Functional Foods: The Consumers the Products and the Evidence*, M. Sadler and M. Saltmarsh, eds. Royal Society of Chemistry, Cambridge, pp. 80–86.

Index

Indexer: Dr Laurence Errington